Microscale Experiments in Organic Chemistry

Second Revised Printing

Pradip K. Bhowmik
Mahin S. Behnia
University of Nevada, Las Vegas

Cover image © Shutterstock, Inc.

Kendall Hunt
publishing company

www.kendallhunt.com
Send all inquiries to:
4050 Westmark Drive
Dubuque, IA 52004-1840

Copyright © 2008, 2013 by Kendall Hunt Publishing Company

ISBN 978-1-4652-3295-3

All rights reserved. No part of this publication may be reproduced, stored in a retrieval system, or transmitted, in any form or by any means, electronic, mechanical, photocopying, recording, or otherwise, without the prior written permission of the copyright owner.

Printed in the United States of America
10 9 8 7 6 5 4 3 2

Table of Contents

Acknowledgements	v
Components of a Microscale Organic Kit	vi
Microscale vs. Macroscale	vii
Procedures for Safety Equipment	vii
Emergency Procedures for CHE: 206	ix
Emergency Procedures for CHE: 219	x
Keeping a Laboratory Notebook	xi
Format for a Lab Notebook	xi
Experiment #1 – Distillation Using the Hickman Still	1
Experiment #2 – Gas Chromatography (Gas-Liquid Chromatography): Analysis of a Mixture	4
Experiment #3 – Thin Layer Chromatography (TLC)	9
Experiment #4 – Recrystallization	14
Experiment #5 – Stereochemistry	19
Experiment #6 – Extraction	23
Experiment #7 – Synthesis of n-butyl bromide (S_N2 Reaction)	29
Experiment #8 and #9 – Elimination: E1 and E2 Reactions	32
Experiment #10 – Addition Polymers	38
Experiment #11 – Diels-Alder Reaction	44
Experiment #12 – Competitive Aromatic Nitration	47
Experiment #13 – Solubility and solution	52
Experiment #14a – Infrared Spectroscopy	59
Experiment #14b – NMR Spectroscopy	69
Experiment #15 – Williamson Ether Synthesis	78
Experiment #16 – Sodium Borohydride Reduction Of ketone	84
Experiment #17 – Grignard Reaction	89

Experiment #18– Fischer Esterification   **96**

Experiment #19 – Dye-Coupling and Diazo-Imaging   **103**

Experiment #20 – Preparation of an α,β-Unsaturated Ketone   **108**

Experiment #21 – Condensation Polymers or Step-Growth Polymers: Nylon and Polyester   **114**

Acknowledgments

This project benefited from the suggestions of reviewers who helped guide the authors through the process. The authors express their sincere thanks and best wishes to Organic Chemistry professors Dr. Lee, Dong-chan associate professor, and Dr. Haesook Han at Chemistry Department, University of Nevada Las Vegas, and Arvin Akoopie, an undergraduate student, who ran several experiments for this project.

The authors, expresses their thanks to those students whose suggestions and course evaluations (241L, 242L, 347L and 348L) have encouraged them to continually assess their instruction so that it could be easily understood. They wish to express their thanks to those students who read and follow these experiments, and provide comments on the further improvement of this project so that their learning becomes efficient and still enjoyable.

Also the support and encouragement of the UNLV Department of Chemistry and the College of Sciences is greatly appreciated.

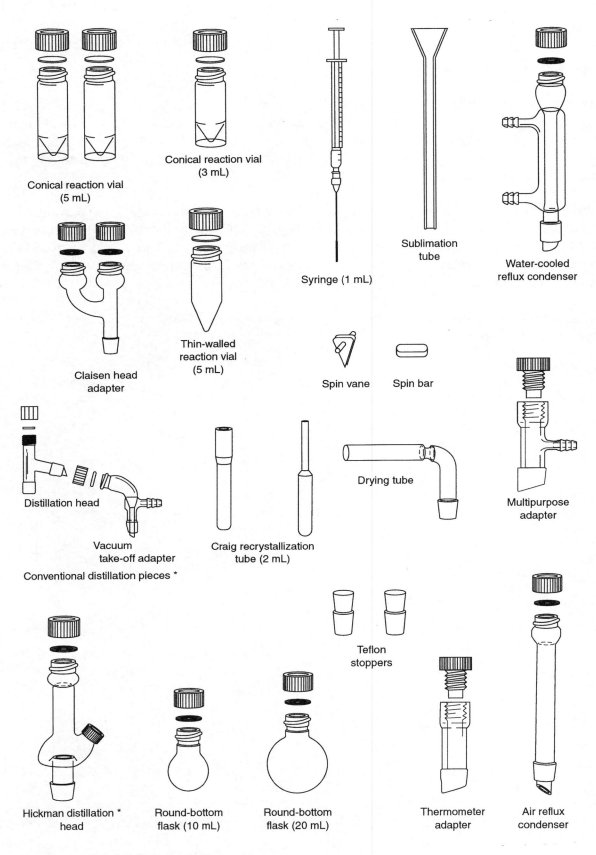

Components of a microscale organic kit.

Microscale vs. Macroscale

Although there are still many precautions that must be taken when utilizing microscale techniques, there are numerous benefits for this type of procedure over those of macroscale procedures. Microscale techniques provide a more efficient method for experimentation as the chemical cost is reduced. For example, the microscale technique utilizes smaller samples thus allowing for less waste of chemical reagents as is normally common in macroscale experimentation. As a safety precaution, there is a reduction in the exposure to hazardous reagents and their vapors as well as limiting the risks involved with working with caustic chemicals. Another key difference when comparing microscale to macroscale is that the microscale techniques can be used to reduce the time necessary to complete the experiment at hand.

Although these safer microscale procedures and techniques are used during our experiments, there still are many safety precautions that must be followed and adhered to. Several examples of these safety procedures and their respective equipment are given below.

Procedures for Safety Equipment
Fire Extinguisher

If it is a small fire in a container, it usually can be extinguished quickly by placing a wire gauze screen with a ceramic fiber center, or possibly, a watch glass, over the mouth of the container. If this method does not take care of the fire and if help from an experienced person is not readily available, then extinguish the fire yourself with a fire extinguisher. This is the procedure you need to follow:

Keep your back to an exit and stand 10 to 20 feet away from the fire. Follow the four-step PASS procedure.

1. Pull the pin: This allows you to discharge the extinguisher (ammonium phosphate).
2. Aim low: Point the nozzle at the base of the fire.
3. Squeeze the lever above the handle: This discharges the extinguishing agent. Releasing the lever will stop the discharge.
4. Sweep from side to side: Moving carefully toward the fire, keep the extinguisher aimed at the base of the fire and sweep back and forth until the flames appear to be out. Watch the fire area. If the fire reignites, repeat the process. Always be sure **the safety officer inspects the fire site** after you have extinguished the fire.

Fire Blanket

If your clothing is on fire, **DO NOT RUN**. Walk purposefully toward the fire blanket or the nearest safety shower station. Running will fan the flames and intensify them. The following is the procedure of how to use fire blanket.

1. Loose the bracket; lift the bracket.
2. Use a swing of your arm to grasp the catch tape. Then, keep rolling it around your body until the blanket detaches from the roller. Next, roll on the floor. Keep doing this until you put out the fire.

Eyewash Station
If any chemical enters your eyes, you need to use the eyewash station immediately. The following is the procedure to follow:

1. Pull down the handle and while holding the eyes open, irrigate your eyes thoroughly for at least 15 minutes.
2. Obtain medical attention for any serious eye injury.

Safety Shower
If your clothing is on fire or you get chemical spills on your body, you need to use the safety shower immediately. If you get CONCENTRATED ACID on your clothing, you need to take off your clothing quickly before taking the safety shower. This is the procedure:

1. Pull the triangle handle to activate the shower.
2. Take this shower for at least 15 minutes.
3. Obtain medical attention in case of serious injury.

First Aid
For minor cuts, burns, and chemical spills on your hand, rinse the affected area with cold water for at least 15 minutes until you no longer feel a burning sensation. If the cut is serious with significant bleeding, apply a cloth compress or tourniquet to control the bleeding. Obtain medical attention for any serious burn or cut. For a minor cut, you can find Band-Aid and alcohol wipe from the kit.

Mercury Spill Kit
If you break a mercury thermometer, you need to use the mercury spill kit. It is on the shelf in CHE 206 and 219. This is the procedure to follow:

1. Wear the designated gloves (obtain from the kit).
2. Obtain sulfur from the kit. Sprinkle some sulfur powder on the spilled mercury (Zinc powder can be substituted for sulfur because it forms amalgam with mercury.)
3. Collect it with the dust pan and brush and put it in the designated plastic bag. The thermometer body should be disposed of in a jar labeled as **BROKEN THERMOMETER. (Under no circumstances should broken thermometers be disposed of in the broken glass container.)**

Waste Disposal
The guidelines for waste disposals are as follows:

1. **Do Not Place Any Liquid Waste or Solid Waste in Sinks.**
2. Dispose of your organic liquid products, unused organic liquid reactants, organic liquid wash, or solvents such as acetone, diethyl ether, hexanes, toluene, etc. in a designated container **"ORGANIC LIQUID WASTES."**
3. Dispose of halogenated organic solvents such as **methylene chloride (dichloromethane), chloroform, and carbon tetrachloride** in a designated container labeled as **"HALOGENATED ORGANIC LIQUID WASTES."**

4. Dispose of your organic solid products or unused organic solids in a container labeled as **"ORGANIC SOLID WASTES."**
5. Strong acids such as hydrochloric, sulfuric, and nitric acids must be neutralized with baking soda before they are disposed of in a container labeled as **"INORGANIC LIQUID WASTES."** (Remember, an explosion happened in CHE: 219 during the summer of 2001 due to disregard of this safety guideline.)
6. Dispose of inorganic solids such as sodium sulfate, magnesium sulfate, silica gel, and alumina in a container labeled as **"INORGANIC SOLID WASTES."**

Specific Safety Procedures

Safety glasses are required and can be purchased from CHE Office (Debbie), Chemical Interaction, or any hardware store. You need to follow the rules listed below in addition to the attached safety rules:
1. In case of a chemical spill, do not clean up; notify the incident to your laboratory instructor.
2. If you break a thermometer, do not clean up; notify the incident to your laboratory instructor.
3. If you get chemical burns or cut your skin, notify the incident to your laboratory instructor.
4. In case of fire, stay calm; notify the incident to your laboratory instructor.
5. If the **fire alarm** goes off, unplug your hot plate and exit the building following the fire escape route.
    a. Meet outside on the east side of the building closest to the **Sidewalk Café**.
    b. Do not return to the building until instructed to do so.
    c. **Failure to observe the safety rules** may result in **dismissal** from the laboratory session and **forfeiture** of the experiment and exercise for that day.

Location of safety and waste Recovery supplies

1. **Acid and Base Spill Cleanup Kit or Sodium Bicarbonate:** North part of CHE 206 and south part of CHE 219.
2. **Fire Extinguisher**: West part of CHE 206 and 219.
3. **Eyewash Station**: South part of CHE 206 and west part of CHE 219.
4. **Safety Shower**: South part of CHE 206 and west part of CHE 219.
5. **Fire Blanket**: East part of CHE 206 and north part of CHE 219.
6. **First Aid Supplies**: West part of CHE 206 and east part of CHE 219 on the shelf.
7. **Mercury Cleanup Kit**: West part of CHE 206 and east part of CHE 219 on the shelf.
8. **Sharp Collector** (for broken glass): South part and under the far end hood of CHE 206 & CHE 219.
9. **Vermiculite** (for organic material adsorption): North part of CHE 206 and south part of CHE 219.

Emergency Procedures for CHE: 206

Nearest Office/Telephone: CHE OFFICE: 895-3510

Evacuation Route: From north exit sign turn right, follow the exit sign, and proceed to the stairs. Meet outside on the east side of the building closest to **Sidewalk Café.**

From east exit sign turn left, follow the exit sign, and proceed to the stairs. Meet outside on the east side of the building closest to **Sidewalk Café.**

Pull-up Station:	Near Dr. Bhowmik's Office: room 207
Fire Alarm:	Opposite of the CHE 206

Fire Extinguisher:	West part of the room
Fire Blanket:	East part of the room
Eyewash Station:	South part of the room
Shower:	South part of the room

Public Safety and Security: 911 or 895-3668
Facilities Maintenance: 895-4357
Env. Health and Safety: 895-4226

Emergency!
In the event of an emergency, you will be notified by your building's alarm and your instructor.

Emergency procedures for CHE: 219

Nearest Office/Telephone: CHE OFFICE: 895-3510

Evacuation Route:	From south exit sign turn left, and proceed to the stairs. Meet outside on the east side of the building closest to **Sidewalk Café.**
	From east exit sign turn right and proceed to the stairs. Meet outside on the east side of the building closest to **Sidewalk café.**
Pull-up Station:	Opposite of the CHE 219 near the Graduate Students' Office
Fire Alarm:	Opposite of the CHE 219 near the Graduate Students' Office

Fire Extinguisher:	South part of the room (next to south door)
Fire Blanket:	North part of the room
Eyewash Station:	North west part of the room
Safety Shower:	North west part of the room

Public Safety and Security: 911 or 895-3668
Facilities Maintenance: 895-4357
Env. Health and Safety: 895-4226

Emergency!
In the event of an emergency, you will be notified by your building's alarm and your instructor.

Keeping a Laboratory Notebook

In order to succeed in the Organic Chemistry Laboratory, you must always be prepared and always keep a detailed record of your experiments. You will learn more, be more efficient, and improve your laboratory experiment in the laboratory manual and when reading about the chemical concepts in your textbook. The laboratory notebook is the quintessential tool of all scientists and should be respected as such. The notebook should be hard covered and bound and all entries made in ink. The following guidelines are provided to help you organize the information in your notebook.

1. <u>Table of Contents</u>: Number each page of your notebook and tabulate all experiments.
2. <u>Date</u>: Each page should have a record of when the entries were made (day and time).
3. <u>Chemical Equations for Reaction(s)</u>: Refer to **"Format for Lab Notebook."**

Format for Lab Notebook

Page#   Date:

Title

1. <u>Goal of the experiment</u>
2. <u>General RXN, Specific RXN, and Mechanisms</u>
3. Physical properties of reagents used in the experiments (in table format)
   > Recall: The "key" players (the reagents involved in stoichiometry).

Reagent	M.F.	Structural Formula	M.W.	M.P.	B.P.

4. Procedure (recipe written in passive voice). Write all your observations. Show all the setups (factor label method) for calculations of the number of moles or mmoles or reagents used. Draw the diagram of specific equipment setup. Show all the calculation setups for experimental yield, theoretical yield, and percentage yield.
5. Results in tabular format.

Results:

	Structural formula of products	Exp. yield	Theo. yield	% yield	Exp. M.P.	Theo. M.P.
A						
B						

6. Characterization of the products: Analysis of FT-IR and ^{1}H NMR (if any) spectra in tabulated format.
7. Discussion of the results. Write your comments regarding any discrepancy between the experimental data and the theoretical data.
8. Postlab questions.

Distillation Using theHickmanStill—Exp#1

Definition
Distillation is a process that involves converting liquid to a vapor state with the aid of heat, and then condensing the vapor back to the liquid state by cooling.

Purpose of Distillation
This technique is used for purification and separation of liquid-liquid mixtures (e.g., methanol and water mixture) and solid-liquid mixtures (e.g., salt and water mixture). If the boiling points of two components in the mixture differ by a large amount (> 100 °C) and if the distillation is carried out carefully, it will be possible to get a fair separation of each component. When the difference between the boiling points of two components is not large (< 100 °C), a fractional distillation must be carried out to result in a fair separation. In both techniques, the initial distillate (condensate) will have the same mole ratio of liquids as the vapor just above the boiling liquid. ***The composition of the vapor will be richer in more volatile components than less volatile ones.*** The closer the boiling points of the components of a liquid mixture, the more difficult they are to be separated by simple distillation.

Types of Distillation:
(a) Simple distillation, (b) Fractional distillation, (c) Steam distillation, (d) Vacuum distillation.

Applications of the Distillation:
This technique has been used around the world by liquor manufacturers, the oil industry, and the chemical industry.

Terminology
- ***Distillate or condensate***: the liquid that has been distilled from the mixture.
- ***Fraction***: a portion of distillate that has been collected at specific temperature ranges.
- ***Volatile component***: a component with a low boiling point.
- ***Boiling point***: a temperature at which the vapor pressure of a liquid becomes equal to the atmospheric pressure.

Guidelines for Distillation
1. You should not fill more than two-thirds of the total capacity of the container.
2. Use a little bit of grease to lubricate the joints to prevent fusion of the joints.
3. Do not heat your mixture strongly.
4. Leave a small opening at the far end of the system *(otherwise the pressure builds up inside a sealed system, the apparatus may explode)*.
5. Using Hickman method, the bulb of the thermometer must be placed in the stem of the Hickman head, just below the well, or it will not read the temperature correctly.
6. Use an aluminum block as a support and heat conductor.
7. Use a spin vane to homogenize heat in the reaction vessel (the conical vial).
8. Do not heat the contents of the vessel to dryness.

NOTE: In this experiment, you will conduct distillation to separate a two component mixture: hexane and toluene, using Hickman Still Functioning as both a condenser and a receiving flask. The collected fractions will be analyzed by GC (Gas Chromatography) to evaluate the accuracy of your distillation technique.

Structures

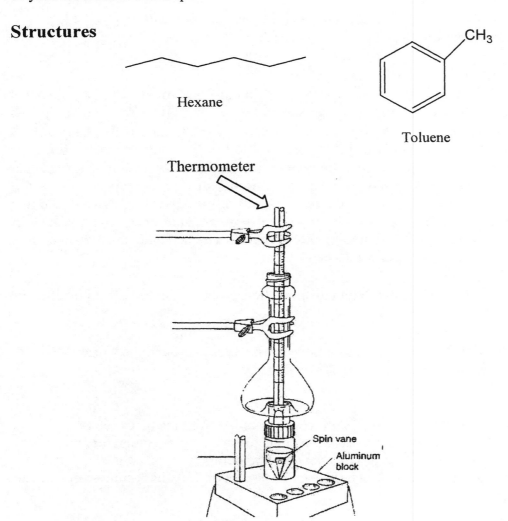

Figure 1: Basic microscale distillation (internal monitoring of temperature)

Procedure
1. Using the "mL" marks on your 5 mL conical vial, measure 1.5 mL of hexane using a Pasteur pipette.
2. Using the "mL" marks on your 3 mL conical vial, measure 1.5 mL of toluene using a Pasteur pipette. Then transfer the contents into the 5 mL conical vial from step 1 in the procedure.
3. Add a spin vane to the 5 mL conical vial.
4. Assemble the apparatus as shown in Figure 1 and make sure that the side arm of your Hickman Still has a cap with a septum.
5. Have your instructor check your setup before you begin the distillation.

6. Set the knob of your hot plate to 3 and adjust the stirrer to the optimum position.
7. Heat the mixture slowly and collect the fractions as follows:
    (a) Collect and remove all condensates up to 69–70 °C in a pre-labeled 3 mL conical vial; this is **Fraction #1**. Quickly cap (Note; the cap should have a septum with no hole in it) the conical vial and seal it tightly with a small piece of aluminum foil and then wrap it with a small piece of parafilm. Leave it in a 150 mL beaker for the next lab period.
    (b) Collect and remove all the condensates up to 104 °C in a pre-labeled 3 mL conical vial. This is the intermediate fraction. Cap and seal it as above in (a).
    (c) At this stage, there should be about 2 mL of liquid remaining in the 5 mL conical vial. This is **Fraction #2**.
    (d) Quickly, and with the help of your lab partner, turn off the hot plate and remove the setup and place it on the bench top. Let the apparatus cool down for about 10 minutes. Remove the Hickman Still and place a cap bearing a septum on the conical vial and seal it as mentioned above in (a).

Reference
Introduction to Organic Laboratory Techniques: A Microscale Approach 3rd ed.; Pavia, Lampman, Kriz, and Engel.

Safety
1. **Take precautions when handling HOT objects.**
2. **If exposure to ANY of the reagents occurs, be sure to wash hands and any other affected areas thoroughly with plenty of soap and water.**
3. **Turn OFF any hot plates that are no longer being used.**
4. **Thoroughly clean (with soap for glassware) and rinse (final rinse of distilled water) ALL glassware and utensils used PRIOR to returning them to your lab drawer.**

Disposal
Dispose of any excess hexane, toluene, and/or distillate fractions in the "Organic Liquid Wastes" receptacle located in the fume hood. Used aluminum and parafilm may be disposed of in a trash or waste basket.

Postlab Questions
1. Draw the graph of temperature versus time at an interval of every two minutes for distillation of a mixture of 1:1 of hexane and toluene and show how effectively the Fractions #1 and #2 are separated compared with an ideal case.
2. What would be your expectations regarding the composition of Fractions #1 and #2? (**HINT**: Are these fractions pure or not?)
3. What are the boiling points of hexane and toluene?
4. Which component will be distilled from the mixture first?

Gas Chromatography (Gas-Liquid Chromatography): Analysis of a Mixture—Exp#2

Definition
Historically, chromatography was used to separate the components of colored compounds (e.g., the separation of pigments in spinach leaves or the separation of pigments in blue ink), but now, modern chromatography is used to separate all types of substances (colored and noncolored materials).

Terminology
Chromatic: related to colors.
Chromatography: the separation of colored materials in a complex mixture.
Modern chromatography: deals with the separation of mixtures containing both colored and noncolored materials.
Analyte: the mixture of compounds to be separated.
Retention time: the period following injection that is required for a compound to pass through the column and emerge as a peak on a chromatogram.
Chromatogram: the signal recorded as current versus time on a graph paper.

Chromatographic Techniques
1. Paper Chromatography
2. Column Chromatography
3. Thin Layer Chromatography (TLC)
4. Gas-Liquid Chromatography (GLC)

Each type involves the interaction of an analyte (e.g., the mixture to be separated) with what is called a mobile phase and a stationary phase.

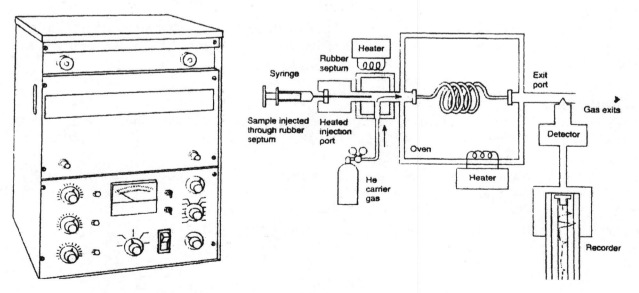

Figure 2: Schematic diagram of a Gas-Liquid Chromatograph

Structure of Gas Chromatograph
1. Mobile phase or moving phase: He or N_2 (carrier gas).
2. Stationary phase: Solid material (the support) coated with a nonvolatile liquid.
3. Column: Copper, stainless steel, or glass with diameter: 1/8 in. (3 mm), ¼ in. (6 mm). Length: 4–12 ft., or up to 50 ft.
4. Packing (the materials inside the column):
    (a) Crushed firebrick
    (b) Teflon beads
    (c) Chromosorb P (diatomaceous earth*) the packing materials are coated with stationary phase.
5. Stationary phase (can be):
    (a) Waxes
    (b) High boiling point hydrocarbons
    (c) Silicone oils (20% DC-200*)
    (d) Polymeric esters, etc.

  * **The cell walls of marine creatures which contain silica.**

Gas Chromatography
A sample is injected into the injection port where it is vaporized and mixed with the carrier gas. The compounds in the injected mixture travel through the column at different rates caused by different interactions that they make with the mobile and stationary phases. Therefore, they emerge from the column at different times. Each component is said to have a unique retention time.

The Principle of Separation
The rate at which different compounds pass through the column is a function of how much time they spend in the vapor phase and how much time they spend in the liquid phase. Therefore, the time that different compounds spend in the vapor phase is primarily a function of their vapor pressures, and the more volatile component arrives at the end of the column first because it spends more time in vapor phase interacting with mobile phase than the less volatile component as illustrated in Figure 3.

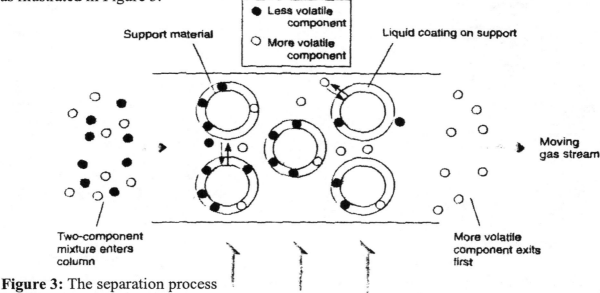

Figure 3: The separation process

Factors Affecting Separation

1. **Temperature of the column**: If it is heated too high, the entire mixture to be separated is flushed through the column at the same rate as the carrier gas. If the temperature of the column is too low, the mixture dissolves in stationary phase and never revaporizes.
2. **The rate of the carrier gas**: At a high flow rate of He gas, the molecules of the sample in the vapor phase cannot equilibrate with those dissolved; this may result in a poor separation between components in the injected mixture. If the rate of the flow is too slow, the bands broaden significantly, leading to a poor resolution. Therefore, the rate of the He gas flow has to be optimum.
3. **The length of column**: The longer the column, the more effective the separation. (Why?)
4. **Choice of the liquid phase used in the column**: *LIKE INTERACTS WITH LIKE;* Silicone oil is used to separate esters, aldehydes, ketones, and halocarbons. Carbowaxes are used to separate alcohols and ethers.
   > Carbowaxes: a polymer of ethylene glycol: **$HO(-CH_2-CH_2-O)_n-CH_2-CH_2-OH$**.

Quantitation

Each peak in the chromatogram (GC graph) represents a component of a mixture. The area under each peak is proportional to the amount, or percent composition, of that component in the mixture.

Determining the percent composition of each component:
1. Mass determination (cut and weigh, old method)
2. Triangulation

**The column we are using in this experiment is packed with Chromosorb P (diatomaceous earth) coated with 20% DC-200 (silicone oil).

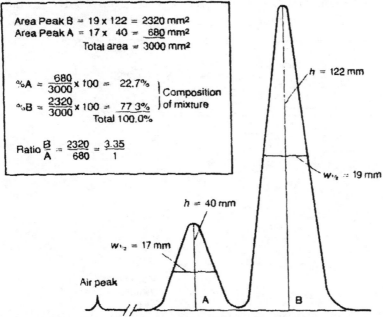

Figure 4: Sample percentage composition calculation by triangulation

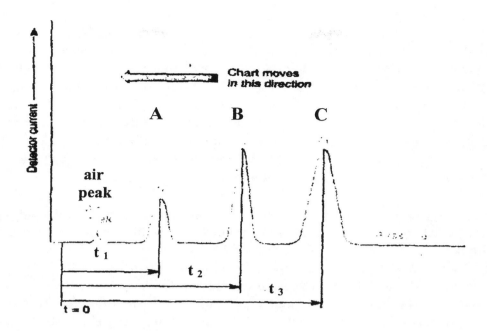

Figure 5: Typical chromatogram indicating retention times (t_1, t_2, t_3) for a three-component mixture

Experimental Procedure—Gas Chromatography

1. Each group will work with one GC assigned by the instructor. (*Use one GC for all fractions. Why?*)
2. In your notebook record all the GC settings (*assigned by your instructor*) and actual temperatures in the table as follows:

Parameters for GC:	GC #1	GC #2	GC #3	GC #4	GC #5
Column setting					
Actual temperature reading for column					
Injector port setting					
Actual temperature reading for injector port					
Detector setting					
Actual temperature reading for detector					
M. A.	100	100	100	100	100

3. Using a 10 µL (µL = microliter) GC syringe, inject 1 µL of the prepared 1:1 mixture of hexane and toluene and 3 µL of air into the injector port of the GC (*GOW-MAC Model 350 with thermal conductivity detector*) located in room 209; your instructor will demonstrate this technique.

4. Calculate the percent composition of this mixture and determine which peak corresponds to hexane and which peak corresponds to toluene. Tape the chromatogram into your notebook. Also calculate the retention time for this mixture using the chart speed of the recorder (typical setting for recorder: 3 cm/min).
5. Using Fraction #1, saved from your distillation experiment, inject 1 µL of Fraction #1 and 3 µL of air into the injector port and follow the instruction as in step 4. Your lab partner will repeat the same procedure for Fraction #2 (the intermediate fraction can be used as a backup for one of these fractions).

- **NOTE: Only one chromatogram per pair. The lab partner who does _NOT_ have a printout of the chromatogram should record the the following data in his/her lab notebook:**
    - **(a) Measured height of each peak of the chromatograms**
    - **(b) Measured width of each peak of the chromatograms**
    - **(c) Measured retention time of each peak of the chromatograms (fraction # 1, #2, prepared mixture 1:1) in her/his lab notebook and show all the calculations in her/his lab notebook. Also, this individual should write the name of his/her lab partner as a reference in her/his lab notebook.**

Reference
Introduction to Organic Laboratory Techniques: A Microscale Approach 3rd ed.; Pavia, Lampman, Kriz, and Engel.

Safety
1. **Take precautions when handling HOT objects.**
2. **If exposure to ANY of the reagents occurs, be sure to wash hands and any other affected areas thoroughly with plenty of soap and water.**
3. **Take precautions when handling SHARP objects (i.e., injection needles).**

Disposal
Dispose of any excess hexane and or toluene in the "Organic Liquid Wastes" receptacle located in the fume hood.
Return **ALL** injection needles to the instructor.

Postlab Questions
1. Using your chromatograms for prepared 1:1 mixture, Fraction #1, and Fraction #2, calculate the percent composition and retention time for each of the above mixtures. Tape all of the chromatograms into your lab notebook (refer to "**NOTE**" at the top of this page). **RECALL: If you have answered this question in the results part of your experiment, do not repeat it again.**
2. Using the results for each fraction, give a comment on the effectiveness of your distillation.

Thin Layer Chromatography (TLC)—Exp#3

Definition
TLC is a method used for the separation of the components of a mixture based on their molecular interactions (adsorption, comprising, e.g., dipole-dipole interaction, coordination, hydrogen bonding, salt formation, etc.) between two phases: mobile and stationary.

Terminology
1. *The stationary phase or adsorbent*: A solid material such as silica-gel $(SiO_2)_x \cdot x\ H_2O$ or alumina $(Al_2O_3)_x \cdot x\ H_2O$ or silicone gum (which is coated on glass or plastic by binders such as gypsum or polyacrylic acid).
2. *The mobile phase:* a single or mixture of solvents which moves up (ascends) the adsorbent by capillary action carrying the analyte.
3. *Eluent*: The mobile phase used in TLC and Column Chromatography.
4. *Retardation factor (R_f)*: The ratio of the distance traveled by the analyte, called the compound front (C_f) to the distance traveled by the eluent, called solvent front (S_f). $R_f = C_f / S_f$
5. *Spotting*: A small circle is generated when the analyte is applied by a capillary tube (one of the steps in TLC technique).
6. *Developing, or running the TLC*: After the sample is spotted on the TLC plate, and the plate is placed in a chamber containing eluent solutions, the solvent ascends the plate and causes the analyte to interact with stationary and mobile phases (this process is called developing or running the TLC plate).
7. *Principle of separation in TLC*: The separation is based on the many equilibrations the analyte experiences between the mobile and stationary phases. In general, the stationary phase is very polar and therefore strongly binds other polar substances (like dissolves like). The moving liquid phase is usually less polar than the adsorbent and most easily dissolves substances that are less polar or even nonpolar. Thus, substances that are the most polar travel slowly upward, or not at all, and nonpolar substances travel more rapidly if the solvent is sufficiently nonpolar.

Factors Affecting Separation
- **Polarity of the analyte:** If a polar compound moves too slowly in a nonpolar solvent, switching to a more polar solvent will cause the compound to move faster. If a nonpolar compound moves too fast during TLC, switching to a less nonpolar solvent will cause the compound to move slower.
- **Polarity of the solvent:** The polarity of the solvent can be varied by mixing miscible solvents to give the desired separation.

Application of TLC
1. To determine the number of components in a mixture.
2. To determine the identity of two substances.
3. To monitor the progress of a reaction.
4. To determine the effectiveness of a purification.
5. To determine the appropriate conditions for a column chromatographic separation.

6. To monitor column chromatography.

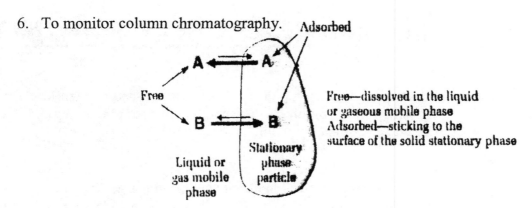

Figure 6: Adsorption equilibrium for molecules between adsorbent and solvent

Elution Order for Some Common Functional Groups with a Silica or Alumina Stationary Phase:

Highest/fastest (elute with nonpolar mobile phase)
- Alkane hydrocarbons
- Alkyl halides (holacarbons)
- Alkenes (olefins)
- Dienes
- Aromatic hydrocarbons
- Aromatic halides
- Ethers
- Esters
- Ketones
- Aldehydes
- Amines
- Alcohols
- Phenols
- Carboxylic acids
- Sulfonic acids

Lowest/slowest (need polar mobile phase to elute)

Common Mobile Phases Listed by Increasing Polarity:
*Suspected carcinogens
- Helium
- Nitrogen
- Pentanes (petroleum ether)
- Hexanes (ligroin)
- Cyclohexane
- Carbon tetrachloride
- Toluene
- Chloroform*
- Dichloromethane (methylene chloride)
- t-butyl methyl ether
- Diethyl ether
- Acetone
- 2-propanol
- Pyridine
- Ethanol
- Methanol
- Water
- Acetic acid

TLC involves the following steps
Step 1: Application of analyte
Step 2: Developing the TLC plate
Step 3: Visualization methods
Step 4: Determining the R_f value in TLC

In the following experiment you will run a TLC plate with six different solid compounds. The six solids are: **naphthalene, 9-fluorenone, benzophenone, 4-methoxyacetophenone, fluorine 4-nitroaniline.**

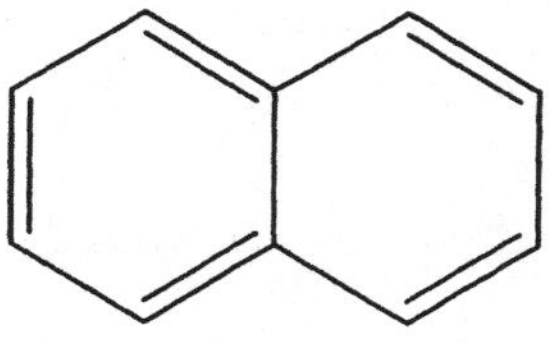

naphthalene 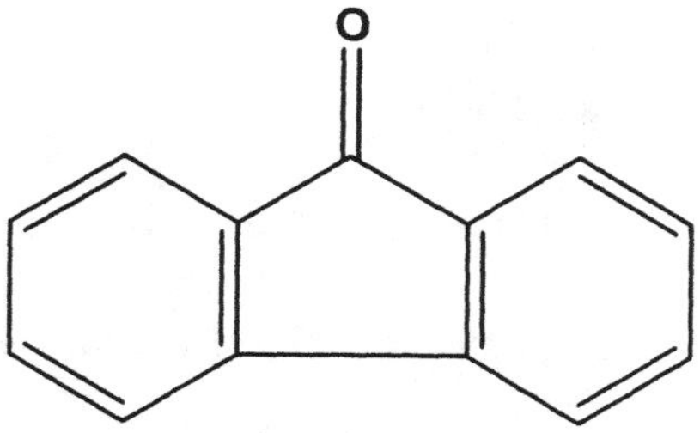 **9-fluorenone** 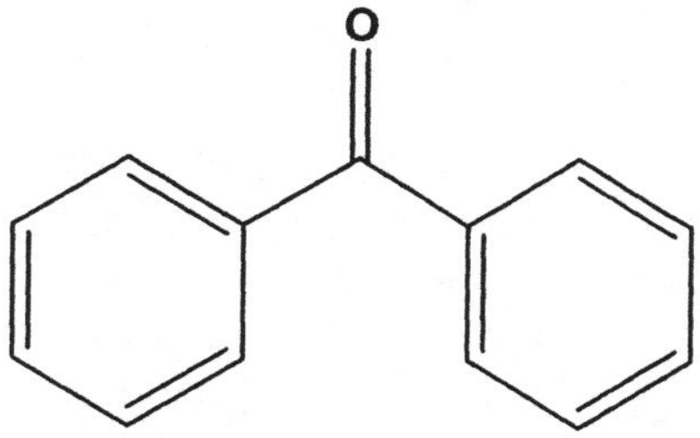 **benzophenone**

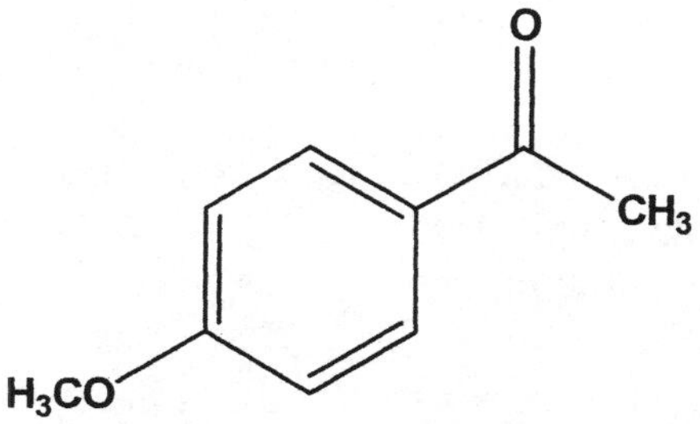

4-methoxyacetophenone 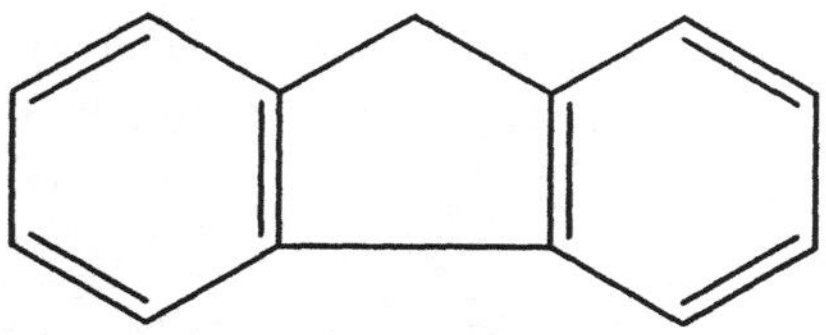 **fluorene** 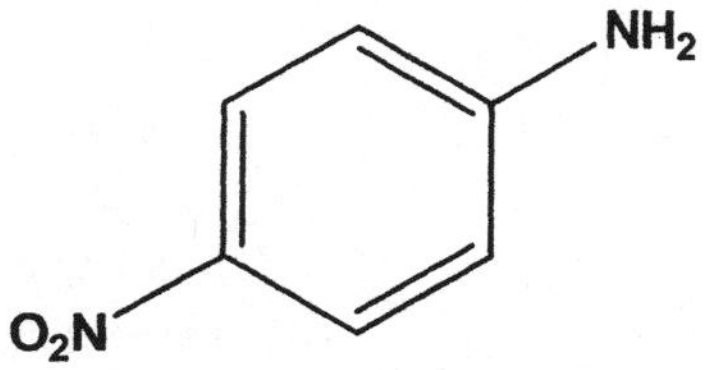 **4-nitroaniline**

Procedure
1. Weigh out 10 mg of each solid and record the amounts weighed (in milligrams and mmols) in your lab notebook.
2. Obtain six clean, dry test tubes and label them for each compound. Transfer each of the compounds to its respectively labeled test tube.
3. Add 1 mL of ethyl acetate (using a graduated cylinder) to each test tube in order to dissolve the solids.
4. Prepare a 10% ethyl acetate/hexane solution (the eluent or mobile phase) by mixing 1 mL of ethyl acetate with 9 mL of hexane using a 10 mL graduated cylinder.
5. Transfer the eluent solution into a 150 mL beaker.
6. Using a paper towel, make a liner (to saturate the atmosphere within the beaker) and place it inside the beaker. Cover the beaker with aluminum foil or a watch glass.

Spotting the TLC Plate
Each group is given *2 and a half* TLC plates. Each lab partner will run a TLC plate using three of the solids. Both lab partners will run one unknown on the *half piece* of TLC plate. The TLC plates must be handled gently as 100 mm thick coating of silica gel can be easily scratched off—***hold the plates by the edges***.
1. *Lightly* draw a faint line 1 cm from the bottom end of the TLC plate using a pencil (why not a pen?) and then three hash marks at intervals of 0.6 cm from each other and from the edges of the TLC plate to guide spotting. ***Refer to the example in Figure 7.***
2. *Lightly* draw a line 1 cm from the top of the plate (this is called the border line for the solvent or the solvent front). ***In addition, write identifying letters at the top of the plate so that you can keep track of the placement of the compound spots.***
3. Assign 1 capillary for each sample tube.

4. Dip the capillary into a sample tube and let the solution move up the pipette (about 2 cm) by capillary action.
5. While holding the capillary tube vertically over the first hash mark on the baseline of the TLC plate, lower the pipette until the tip just touches the hash on the adsorbent.
6. Withdraw the capillary when the spot is about 2 mm in diameter, then let the solvent evaporate.
7. Next, repeat the application of the same sample ***three more times*** in the ***same spot*** (to ensure that sufficient material is present).
8. Repeat steps 5, 6, and 7 using a different sample on the next available hash mark and using a fresh capillary tube. Repeat this step for the rest of the samples on their designated areas.

Development of TLC Plate
1. Place the plate into the development chamber so that the plastic support faces the wall of the beaker. In addition, the plate should not come in contact with the liner. Also make sure that the origin spots are ***NOT*** below the solvent level in the chamber; otherwise, the analytes will be washed off of the plate and lost.
2. Monitor the advancement of the solvent (by capillary action) from the back of the plastic support.
3. Once the solvent reaches the solvent front, remove the plate using tweezers *(the proper location of the solvent front is important for R_f calculations)* and place it on a paper towel (silicagel face up) and let the solvent evaporate.

Visualization

If the organic molecules are colored, such as with dyes, inks, or indicators, the visualizing the separated spots is easy. However, because most organic compounds are colorless, this is rarely the case. For most compounds, a UV light works well for observing the separated spots. TLC plates normally contain a fluorescent indicator that makes them glow green under UV light of wavelength 254 nm. Compounds that absorb UV light at this wavelength will quench the green fluorescence, yielding dark purple or bluish spots on the plate.

> **Visualization Procedure**

1. Turn on the UV lamp.
2. Press the button for short wavelength.
3. Hold the UV lamp over the spotted TLC plate.
4. Lightly circle the spots with a pencil so that you will have a permanent record of their location for R_f calculations.

R_f Calculations
Conditions include:
1. Solvent system
2. Absorbent
3. Thickness of the adsorbent layer
4. Relative amount of material spotted

Under an established set of such conditions, a given compound always travels a fixed distance relative to the distance the solvent front travels. It is expressed as:

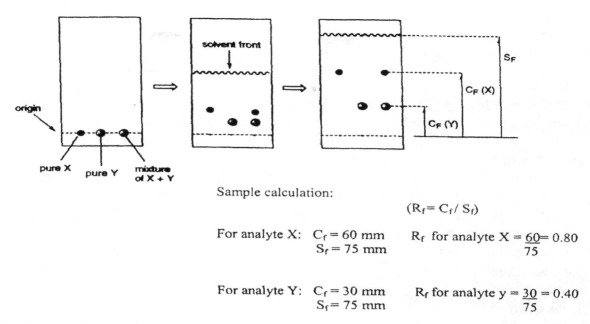

Figure 7: R_f sample calculation for TLC

Results
Draw pictures of your TLC plates in your lab notebook (refer to Figure 7).
The calculated R_f values for each of the analytes should be presented in your lab notebook.

References
Introduction to Organic Laboratory Techniques: A Microscale Approach 3rd ed.; Pavia, Lampman, Kriz, and Engel.

Safety
1. Take precautions when handling the TLC plates.
2. **Be sure to wash hands and any other affected areas thoroughly with plenty of soap and water.**
3. **Thoroughly clean (with soap for glassware) and rinse (final rinse of distilled water) ALL glassware and utensils used PRIOR to returning them to your lab drawer.**

Disposal
Dispose of any excess organic reagents in the "Organic Liquid Wastes" receptacle.
Dispose of your TLC plates in a zipper closure bag labeled "TLC plates contaminated with organic compounds."

Postlab Questions
1. Based on the R_f values determined for each of the six compounds, which compound is the most polar? Which is the least polar?
2. If you used a solution of 90% ethyl acetate/hexane instead of a 10% ethyl acetate/hexane to elute your TLC plates, what would you expect to see? Would the R_f values increase or decrease for each of the six compounds?

Recrystallization—Exp#4

General Purpose of Recrystallization
This technique is used for purification of crude organic products.

Background
The products resulting from organic reactions are seldom pure because unreacted material and by products may be trapped within the crystal lattice or upon the surface of the solid crude product. Washing the crystals with cold solvent can remove adsorbed impurities from the surface, but this process cannot remove the trapped (occluded) impurities. To remove these we need to redissolve the solid in hot solvent, filter off any insoluble impurities, *and then cool the solution to let the material crystallize again.*

Definition
Organic solids are generally more soluble in hot solvent than in a comparable volume of cold solvent. In recrystallization, a saturated solution is formed by carefully adding an amount of hot solvent just necessary to dissolve a given amount of solid. As the solution cools, the solubility of the solid decreases and the solid crystallizes. Unavoidably, some of the solid remains dissolved in the cold solvent, so that not all of the crystals dissolved originally are recovered.

✓ **To remove impurity from the crude product, keep in mind the following:**
1. The solid and the impurities should have differing solubilities in a certain solvent. Impurities that remain undissolved in the solvent can be removed by hot gravity filtration of the solution prior to cooling.
2. Impurities that remain dissolved will remain in solution after the solid crystallizes by quickly or gradually cooling the saturated solution. The impurities can be separated from the product by filtering the saturated solution.

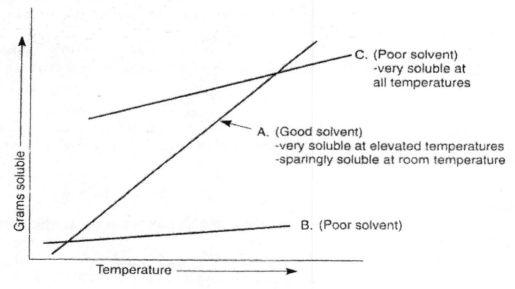

Figure 8: Graph of solubility vs. temperature

Recrystallization Process
1. To choose a suitable solvent that dissolves a small amount of crude product being purified at room temperature, but only a large amount at an elevated temperature (refer to the solubility graph).
2. To dissolve the crude product in a minimum amount of boiling solvent.
3. To allow the saturated solution to cool and to deposit crystals of the compound.
4. To filter the crystals from the solution (mother liquor) as shown in Figure 10.

Procedure [recrystallization of small amount of solid (~0.1 g)]—microscale method

1. Fill a 100 mL beaker about half-full of water and heat to near boiling (~70 °C) on a hot plate.
2. Weigh out 0.78 mmol of naphthalene to the nearest 0.1 g **(i.e., 0.095–0.105 g)** using weighing paper and record the exact amount weighed in your notebook.
3. Transfer the naphthalene to a clean test tube portion of a Craig tube. Add a few drops of a 95% ethanol to the naphthalene in the Craig tube.
4. Place the Craig tube in a hot water bath, hot sand bath, or steam bath **ONLY**! THINK: *Why not use a hot plate?*
5. Continue to add ethanol until the naphthalene just dissolves.
6. The total amount of added ethanol should not exceed more than 1 mL. **WHY?** (**NOTE:** *If the total amount of solvent exceeds 1 mL, the excess solvent should be evaporated.*)
7. Remove the Craig tube from the water bath, cover the top with aluminum foil and place it in a 25 mL Erlenmeyer flask to cool. Put a 150 mL beaker upside-down over the Erlenmeyer flask to allow for a slower, more homogeneous cooling.
8. Get the Teflon plug from your micro kit and wrap (by twisting) the end of an 8-inch piece of nichrome wire around the neck of the Teflon plug (your instructor will demonstrate this to you). Insert the plug into the Craig tube.
9. Place a centrifuge tube over the top of the entire assembly while holding and pulling the wire firmly, invert the centrifuge tube and wrap the extended portion over the lip of the centrifuge tube. Also, follow the **safety rules** associated with the **centrifuge.**
10. Place the assembly in the centrifuge. Make sure the centrifuge is counterbalanced either by another student's apparatus, or by a centrifuge tube filled with enough water to be the same weight as your sample.
11. Turn on the centrifuge for about 1.5 minutes. When the spinning stops, remove the apparatus from the centrifuge.
12. Lift the Craig tube assembly out of the centrifuge tube by lifting the wire.
13. Remove the filter plug and scrape the crystals out onto a clean, dry, pre-weighed watch glass. If there are crystals still remaining at the bottom of Craig tube, repeat the centrifugation for a longer time (~2 minutes).
14. Let the crystals air-dry for at least 30 minutes. When they are completely dry, determine the mass and calculate the percent recovery of your product.
15. Determine the melting points for both the impure naphthalene and the recrystallized naphthalene. Compare these values to that of the theoretical value for naphthalene in literature.
16. Repeat the recrystallization procedure using hexane instead of 95% ethanol.

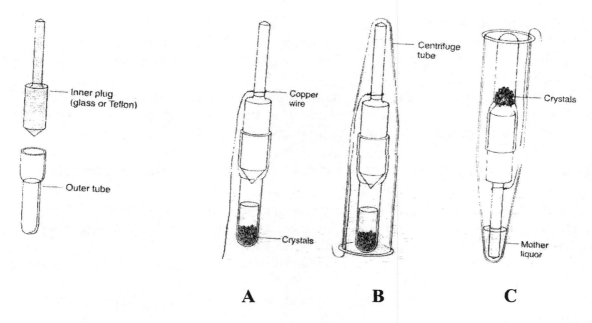

Figure 9: Craig tube (2 mL)

Figure 10: Separation with a Craig tube showing the steps: A, B, and C

Show your results in a tabular format as shown below:

Mass recovery	% Recovery	Melting point of crude or impure naphthalene:	Melting point of the naphthalene after recrystallization
Hexane:			
Ethanol:			

Melting Point Measurement

Most organic compounds have melting points below 300 °C. The melting point of a pure solid compound is a physical property that can be measured as a method of identification. Melting points of pure compounds are recorded in handbooks of physical data, such as the Handbook of Chemistry and Physics (CRC). The reported melting point is that temperature at which solid and liquid phases exist in equilibrium. You should report the melting point of your product or unknown in a range, which starts from the temperature at which the first drop of liquid appears and ends at the temperature at which the entire sample turns into liquid. The impurity generally lowers and/or broadens the melting point.

Melting Behavior of Solids

In many cases, the crystals will soften and shrink immediately before melting. The crystals may appear to "sweat" as traces of solvent or air bubbles are released. These are normal occurrences, but shouldn't be considered as "melting." The melting point is measured beginning at the time the first free drop of liquid is seen. Sometimes compounds decompose rather than melt. Decompositions usually involve changes such as darkening or gas evolution. Decomposition may occur at or even below the melting point for compounds that are thermally labile. Some compounds can change directly from a solid to a gas without passing through a liquid phase, this is called sublimation. In order to measure a melting point (more accurately, sublimation point) for such compounds, a sealed evacuated melting point capillary tube must be used.

How to Determine a Melting Point
Preparing the Sample
Use 1–2 mg of a dry and ground solid sample and load it into the capillary tube by pushing the open end of the capillary down onto the sample and tapping on the solid sample. Then invert the tube. Gently tap the bottom of the tube on the bench or drop the tube through a short 2-foot piece of glass tubing; this causes the sample to pack more tightly and gives a more accurate melting point [Figures 11 (a) and (b)]. The height of sample in the capillary tube should be about 1–2 mm. If you load too much sample into the capillary tube, it will cause the measured melting point to be wide and slightly high.

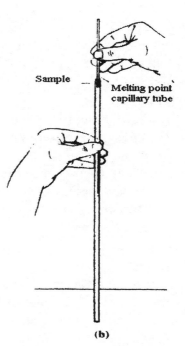

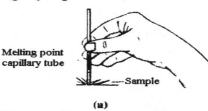

Figure 11: Loading a sample into a melting point capillary tube

Setting the Heating Rate
If the melting point is unknown, heat the sample rapidly to establish an approximate melting point. Turn off the apparatus as soon as the compound melts and note the temperature. Let the temperature drop until it is approximately 20 °C below the observed melting point and repeat the melting point determination with a new sample. Heat the sample rapidly to within 20 °C of the known melting point. Then, slow to 1–2 °C per minute. Heating too rapidly results in inaccurate, usually wider, melting point measurements. An appropriate heating rate can also be determined by referring to the heating-rate curve that often accompanies the melting point apparatus. Record the melting point range as the temperature at which melting starts to the temperature when all solid is converted to liquid. Remember that shrinking, sagging, color change, texture changes, and sweating are *not* indications of melting. When the sample has melted, turn off the melting point apparatus and remove the capillary tube. Discard the used capillary tubes in the glass disposal container.

Procedure (for Operating Melting Point Devices)
1. Insert the capillary tube in the sample compartment. (**NOTE:** You can run two samples simultaneously.)
2. Turn the power switch on.
3. Set the power level at 3 to obtain the desired heating rate.
4. Read the temperature when you observe a drop of liquid in your capillary tube and record it in your lab notebook (this is your initial temperature).

5. Read the temperature when the whole sample turns into liquid (this is your final temperature) and record it in your lab notebook. Remember, always present your measured m.p. in ranges (e.g., m.p. for benzoic acid: 122–124 °C).

References
Introduction to Organic Laboratory Techniques: A Microscale Approach 3rd ed.; Pavia, Lampman, Kriz, and Engel.

Safety
1. **Take precaution when handling HOT objects.**
2. **Turn OFF any burners, hot plates, and/or any melting point devices when not in use.**
3. **Clean (using soap for glassware) and rinse (using distilled water for final rinse) ALL glassware and utensils PRIOR to returning them to your lab drawer.**
4. **If exposure to ANY of the reagents occurs, be sure to wash hands and any other affected areas thoroughly with plenty of soap and water.**

Disposal
Dispose of any capillaries and other disposable glassware in the "broken glass" bin.
Dispose of any solid reagents and/or products in the "Organic Solid Wastes" receptacle.
Dispose of any liquid reagents and/or products in the "Organic Liquid Wastes" receptacle.

Postlab Questions
1. Why is it important to allow for slow cooling?
2. What does it mean when a compound melts over a broad range of temperatures?
3. Why is it important that the total volume of ethanol used in the recrystallization be less than 1 mL?
4. Which solvent was more effective for the recrystallization of naphthalene, ethanol or hexane? (Full explanation is required for full credit!)

Stereochemistry—Exp#5

Background

Stereochemistry is the aspect of chemistry that deals with molecular structure in three dimensions. **Isomer** is a general term used to describe compounds that have the same molecular formula but differ in the arrangement of their atoms. There are several different types of isomers as shown in the diagram below. **Constitutional isomers** are also called structural isomers and they are molecules that differ from each other only in the way the atoms are connected to each other. For example, *n*-butane and *iso*-butane are constitutional isomers. The "eclipsed" and "staggered" forms of ethane are considered to be **conformational isomers** (a.k.a. **conformers**) because they are related to each other through rotation about single bonds (such as C-C, C-O, and C-N) within the molecule. **Geometric isomers** are molecules that differ from each other in the relative arrangement of atoms around double bonds or rings. For example, *cis*-2-pentene and *trans*-2-pentene are geometric isomers.

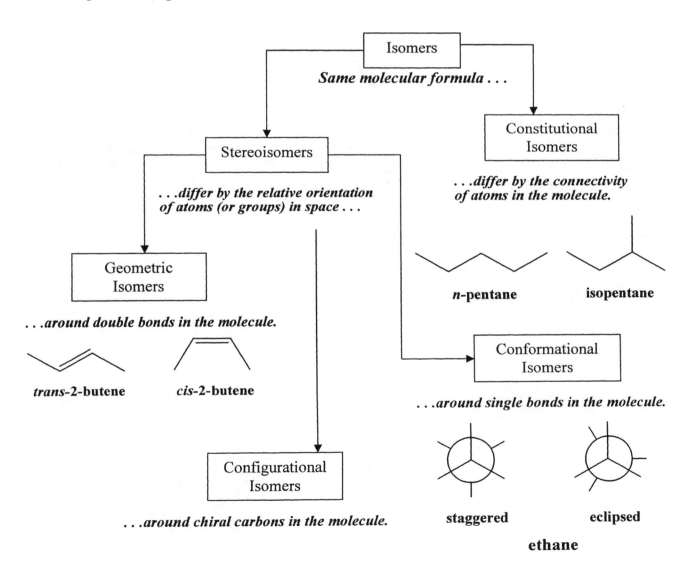

19

Configurational isomers (a type of stereoisomer) are molecules that differ from each other by the relative orientation of atoms, or functional groups, around ***chiral carbon*** atoms. Chiral essentially means not symmetrical. A ***chiral carbon*** is any carbon atom that has four ***different atoms*** (or groups) attached to it. There are a maximum of 2^n ***configurational isomers*** for any given compound, where ***n = the number of chiral carbons in the molecule***. Hence, if you have a molecule with one chiral carbon, there will be no more than two configurational isomers for that compound. If this pair of configurational isomers are mirror images of each other, and nonsuperimposable, they are called ***enantiomers.***

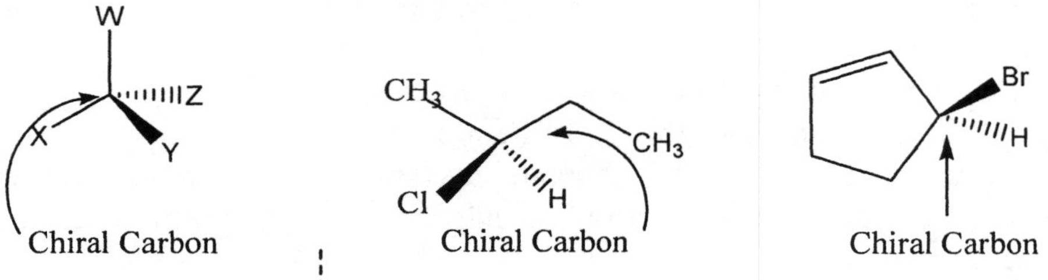

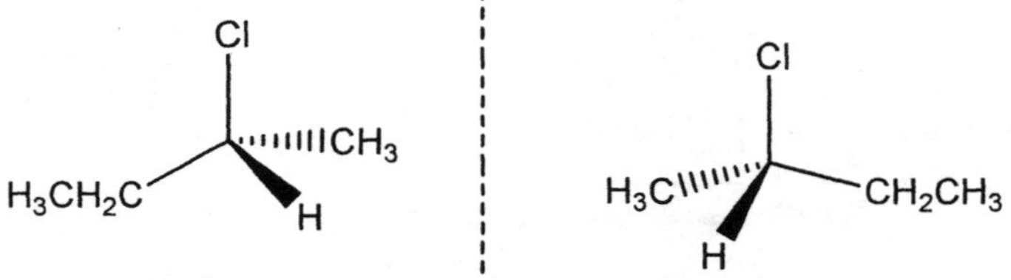

The arrangement of atoms around a chiral carbon that characterizes a particular stereoisomer is called its ***configuration.*** Each enantiomer in a pair is "named" according to this configuration. To determine the configuration of a given chiral carbon, the following three steps must be applied. *(You are required to consult your organic chemistry textbook and lecture notes for more detailed discussions of these concepts. Please bring your textbook to lab.)*

STEP 1: Assign priorities to the four different atoms (or groups) around the chiral carbon according to the **Cahn-Ingold-Prelog** scheme. The highest priority group is assigned the numeral 1 and the lowest priority group is assigned the numeral 4.

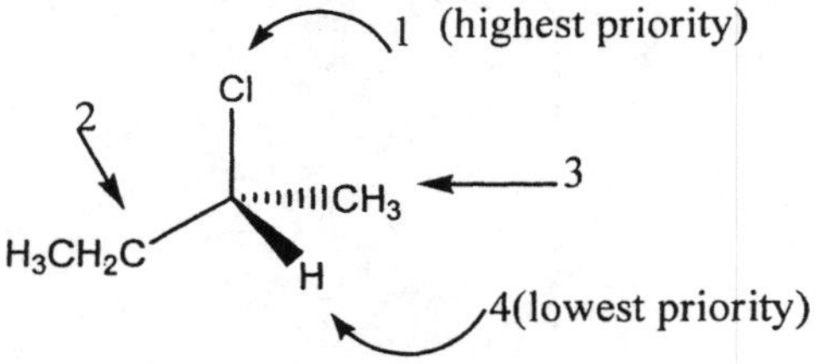

STEP 2: Redraw the molecule, if necessary, so that the lowest priority group is behind the plane of paper. *USE A MODEL!!!*

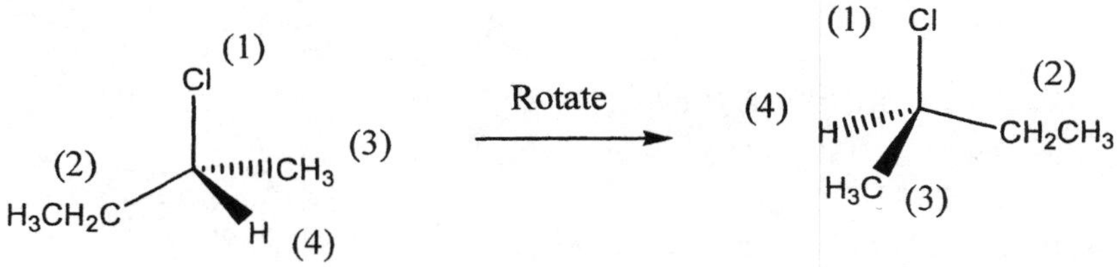

STEP 3: Determine if the relative order of the remaining three groups increases in a clockwise fashion or in a counterclockwise fashion. If clockwise, the configuration is "R"; if counterclockwise, the configuration is "S."

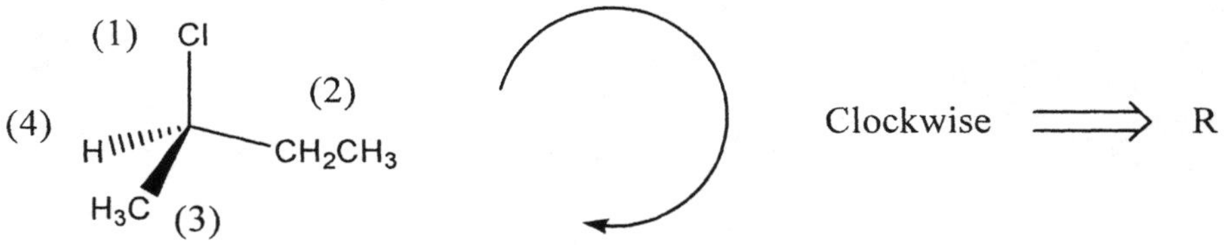

Clockwise ⟹ R

Note that the mirror image isomer of this molecule has the S configuration. For a molecule with one chiral carbon, such as 2-chlorobutane, the R and S isomers are always enantiomers.

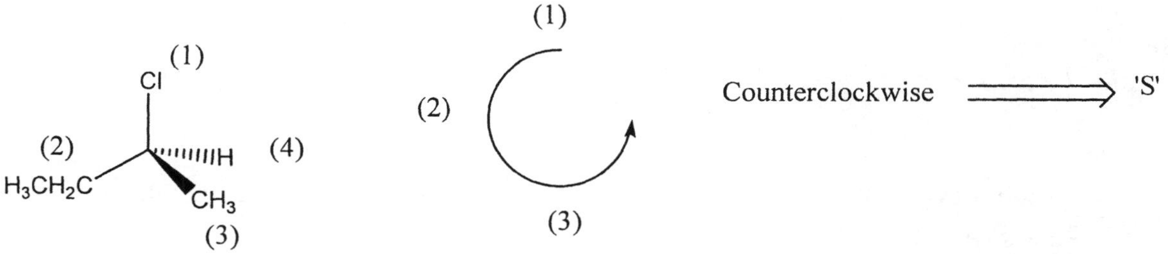

Counterclockwise ⟹ 'S'

Many naturally occurring organic compounds contain multiple chiral carbons and therefore will have several configurational isomers. These isomers will not all be related as enantiomers. ***Diastereomers*** are configurational isomers of a molecule that are *not* enantiomers and *not* mirror images. For example, consider the molecule 2-bromo-3-chlorobutane. This compound has 2 chiral carbons and therefore exists as four possible configurational isomers.

21

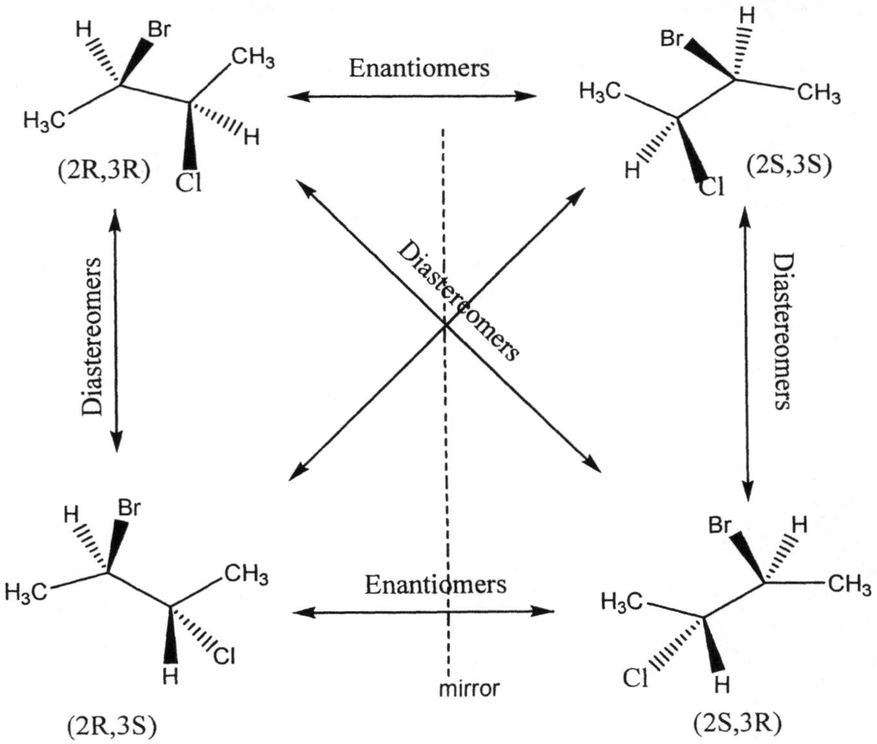

One other possibility exists for molecules that have more than one chiral carbon. Configurational isomers that have two or more chiral centers and an *internal* plane of symmetry are called *meso* compounds. (2R,3S)-1,2-dibromo-1,2-difluoroethane is a meso compound.

(2R,3S)

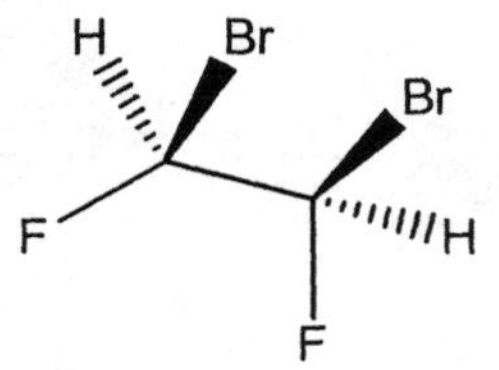

Meso: 2 chiral carbons and an internal plane of symmetry

In this experiment you will see and use models to learn the concepts of stereochemistry.

References
Organic Chemistry 7th ed.; John McMurry.

Safety
Use your model sets productively, abstain from any horseplay (i.e.; throwing, jabbing, or any other misuse of the pieces).

Disposal
There are no disposal procedure guidelines for this experiment.

Postlab Questions
There are no postlab questions for this experiment—complete the handout only.

Extraction—Exp#6

Background
Extraction is one of the most common methods of separating an organic product from a reaction mixture or isolating a natural product from a plant.

Types of Extraction
- liquid-liquid
- solid-liquid

In liquid-liquid extraction the analyte is usually distributed between two different liquid phases: the organic solvent and water. The efficiency of extraction will depend upon the solubility of the compound in the two solvents. The ratio of solubilities is called the distribution coefficient (K_D). By usual convention, the distribution coefficient is equal to the solubility of a compound in an organic solvent divided by the solubility in water.

$$K_D = \frac{\text{solubility in organic solvent (g/100 mL)}}{\text{solubility in water (g/100 mL)}}$$

The magnitude of K_D gives an indication of the efficiency of extraction: the larger the value of K_D, the more efficient the extraction. If K_D is 1, equal amounts of the compound will dissolve in each of the phases. If K_D is much smaller than 1, the compound is more soluble in water than in organic solvent. Therefore, it is better to search for a different organic solvent in which the compound is more soluble in order to perform a more complete separation by liquid-liquid extraction.

Purpose of Extraction
Purification and separation of components of a mixture based upon their acidity, basicity, and solubility properties.

To serve as a good solvent for extraction of an organic compound, the solvent should:
- have high solubility for the organic compound.
- be immiscible with the other solvent (usually water).
- have a relativity low boiling point so as to be easily removed from the compound after extraction.

Typical solvents used in extraction are: diethyl ether, dichloromethane, hexane, ethyl acetate, and petroleum ether.

In the following experiment you will separate a two component mixture: benzoic acid (acidic component) and 9-fluorenone (neutral component) using extraction technique.

The Extraction Technique
The following diagram depicts a general separation via extraction technique. Although a three-component mixture is used in the following example, the mixture you will be analyzing may only have two components. The basic concept of the extraction procedure itself is relatively

simple. The mixture to be analyzed is first separated into organic and aqueous layers using diluted NaOH and dichloromethane. The aqueous layer, upon addition of concentrated HCl, becomes the acidic component of the mixture. The organic layer from this first separation is then separated into a second formation of organic and aqueous layers using a dilute HCl solution. The second organic layer formed can be further manipulated to yield the neutral component of the mixture as the second aqueous layer formed is neutralized to form the basic component of the original mixture.

Separation of a Three-Component Mixture Using Extraction Technique

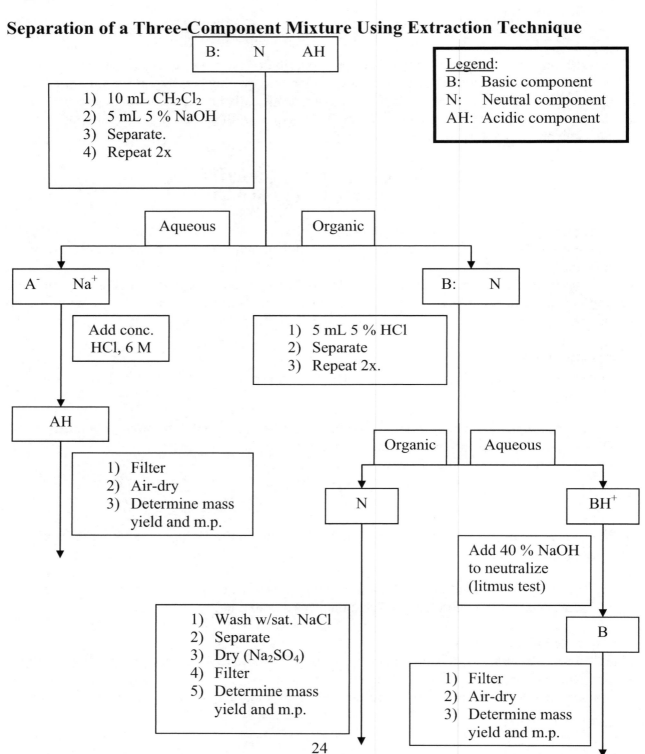

The major key reactions are as follows:

A mixture of benzoic acid and 9-fluorenone in diethyl ether

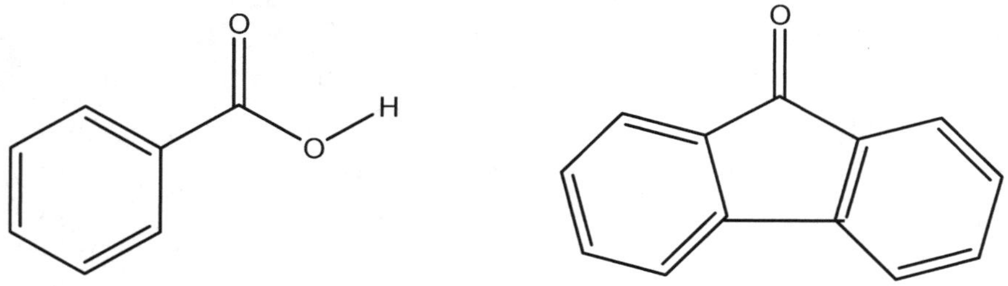

benzoic acid

9-fluorenone

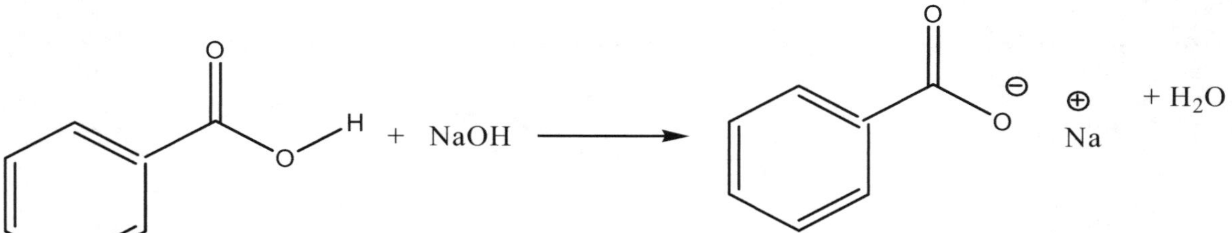

organic phase containing benzoic acid + NaOH → aqueous phase containing soluble salt + H_2O

The major key steps are as follows:

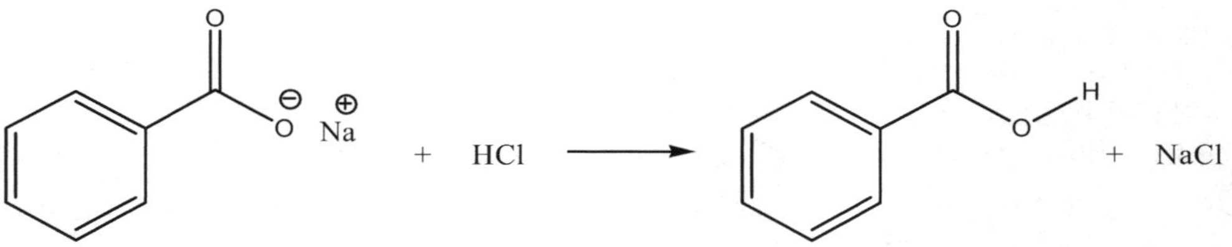

sodium benzoate + HCl → benzoic acid (precipitate out in aqueous phase) + NaCl

Step 1: Dissolve the mixture in diethyl ether.
Step 2: Add extraction solvent: sodium hyroxide to the above solution and stir or shake the solution in order to transfer the benzoic acid to the aqueous layer.
Step 3: Separate the layers.
Step 4: Add hydrochloric acid (HCl) to aqueous layer to convert sodium benzoate to benzoic acid. The benzoic acid will precipitate out from the aqueous solution.

Procedure
1. Weigh out 0.100 g each of benzoic acid and 9-fluorenone (using weighing paper) and record the exact amount weighed (in grams and mmols) in your lab notebook.
2. Transfer each sample into a 5.0 mL conical vial containing a spin vane.
3. Add 2.0 mL of diethyl ether to the vial and stir to dissolve the compounds.
4. Then run TLC on the mixture using the instruction given by your TA. (hint: you need to dilute your sample several times to get good results for TLC).
   NOTE: *In order to prevent the solution from being splashed, do not stir vigorously.*

5. Add 1.0 mL of 5% aqueous sodium hydroxide (NaOH) solution to the vial. Stir the mixture for 1 minute so that the two layers mix into each other and form a tornado-like solution.
6. Let the layers separate in the vial. Carefully pipette the lower layer (squeeze the bulb to remove the air) by inserting the tip of the pipette into the bottom layer of the conical vial and slowly releasing the bulb. Carefully draw the liquid up into the pipette as shown in Figure 12.

A. The ether solution contains the two component mixture.

B. Sodium hydroxide solution is used to extract the acidic component from the organic phase (ether layer).

C. The lower aqueous layer is removed from the organic phase.

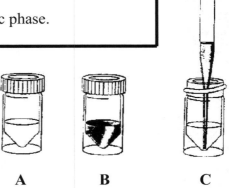

A   B   C

Figure 12: Extraction of an organic solution using sodium hydroxide solution.

7. Transfer the aqueous layer from the pipette to a 25 mL clean Erlenmeyer flask and save it—label it "aqueous."
8. Add another 1.0 mL of 5% aqueous NaOH solution and repeat the extraction procedure. Once again, using the same pipette draw off the aqueous layer and add it to the 25 mL Erlenmeyer flask containing the first basic extract.
9. Add 1.0 mL of saturated NaCl to the vial containing organic layer (stir the solution for 1 minute so that the two layers mix into each other).
10. Draw off the aqueous layer (which layer?) by pipette into a small beaker and save it and label it "saturated salt solution."

Procedure (Drying the Organic Solution—Removing the Water)

1. Remove the spin vane using a spatula.
2. Add a small amount (about a half of a tip of a spatula) of anhydrous sodium sulfate (drying agent) to the organic solution in the 5 mL vial and swirl and twirl (using a spatula).
3. Repeat the above step if the drying agent clumps (the solution should look clear and the additional drying agent should stay free-flowing).
4. Swirl and let the solution stand for 5 minutes.

5. Prepare a Pasteur filter pipette that has been fitted with a cotton plug and clamp it to a ring stand (using a thermometer clamp) above a preweighed 5 mL conical vial containing a piece of boiling chip.
6. Prepare a filter tip pipette (your instructor will demonstrate this technique). Then, squeeze the bulb to remove the air. Insert the pipette into the bottom of the conical vial. Release the bulb slowly and draw the liquid up into the pipette. **NOTE:** *Avoid drawing up the drying agent!*
7. Drain the liquid into the filter pipette (from step 5—drying the organic solution).
8. Collect the effluent into a 5 mL vial.
9. Rinse the drying agent and also the wall of the conical vial with additional small portions of solvent (0.5-1 mL of warm ether) and swirl it for a few minutes and add it to the filter pipette (using the same filter tip pipette).
10. Repeat the rinsing one more time if the drying agent bears a tinge yellow color.
11. Rinse the cotton plug with a several drops of warm diethyl ether and collect the rinse in the 5 mL vial. **NOTE:** *If the tip of filter pipette clogs with yellow solid, rinse the tip with a small amount of warm diethyl ether and collect it in the 5 mL vial.*
12. Discard the drying agent in the "Inorganic Solid Waste" container.

Isolation of the Neutral Compound

1. Fill a 100 mL beaker about half-full with water and heat to ~ 45 °C.
2. Add a preweighed boiling chip to the 5 mL vial and then place it in the warm water bath (**DO NOT** *allow the bath to exceed two-thirds of the height of the vial*) until all of the diethyl ether has been evaporated. **NOTE:** *Take precaution when handling diethyl ether. It is flammable and toxic in large doses!!*
3. Let the vial cool down and then determine the mass of your solid 9-fluorenone and calculate the percent recovery. Also, run TLC on the recovered 9-fluorenone versus reference materials (commercial 9-fluorenone and benzoic acid)

Isolation of the Organic Acid from the Aqueous Basic

1. Put the 25 mL Erlenmeyer flask *(containing the basic extracts and labeled as "aqueous")* in a 250 mL beaker containing ice.
2. Measure 5 mL of a 6.0 M hydrochloric acid using a graduated cylinder.
3. Add acid dropwise to the Erlenmeyer flask (step 1) while swirling the contents in the ice water mixture. Keep adding hydrochloric acid to the Erlenmeyer flask until the benzoic acid starts to precipitate out from the solution as a white precipitate.
4. Run a litmus test to make sure the precipitation is complete. Dip a glass rod in the suspension and rub it against a blue litmus strip (if the color turns to red, it means that neutralization is complete).
5. Stir the suspension and let it stand for 15 minutes.
6. Set up the vacuum filtration apparatus using a Hirsch funnel, 1.0 cm filter paper disk, and a side-arm test tube (your instructor will demonstrate this technique).
7. Wet the filter paper with a small amount of water and turn on the vacuum to medium strength.
8. Swirl rapidly the Erlenmeyer flask containing precipitated benzoic acid and pour the suspension into the Hirsch funnel.

9. Leave the vacuum on *(turn to high strength)* to draw air through the crystals until no water drips from the stem of Hirsch funnel.
10. Leave the vacuum on for a few more minutes to draw air through the crystals.
11. Transfer the crystals to a clean, dry, preweighed watch glass. Break and spread the precipitate into small particles with spatula and let them dry for at least 30 minutes.
12. When they are completely dry, determine the mass of your recovery, calculate the percent recovery, and **run TLC on the recovered benzoic acid against the mixture of your reference materials (co-spot).**

References.
Introduction to Organic Laboratory Techniques: A Microscale Approach 3rd ed.; Pavia, Lampman, Kriz, and Engel.

Safety
1. **Take precaution when <u>handling HOT objects</u>.**
2. **<u>Turn OFF any</u> burners and/or hot plates when not in use.**
3. **Clean (using soap for glassware) and rinse (using distilled water for final rinse) <u>ALL</u> glassware and utensils <u>PRIOR to returning</u> them to your lab drawer.**
4. **If exposure to <u>ANY</u> of the reagents occurs, be sure to wash hands and any other affected areas thoroughly with plenty of soap and water.**
5. **6 M HCl is corrosive. So, prevent contact with eyes, skin, or clothing.**
6. **Ether is highly flammable: No flames should be allowed in the laboratory when this experiment is being performed.**

Disposal
Dispose of your benzoic acid in the beaker labeled as "receptacle for benzoic acid" and dispose of 9-fluorenone in the beaker labeled as "receptacle for 9-fluorenone." If these receptacles are not available, dispose of these solid reagents in the "Organic Solid Wastes" receptacle.
Dispose of the disposable glassware in the "broken glass" bin.

Postlab Questions
1. You have been given a mixture (1:1) of the two compounds shown below, diethyl ether, aqueous $NaHCO_3$, aqueous NaOH, and aqueous HCl. Using a flowchart, outline your procedure and explain how to separate, isolate, and purify these two compounds.

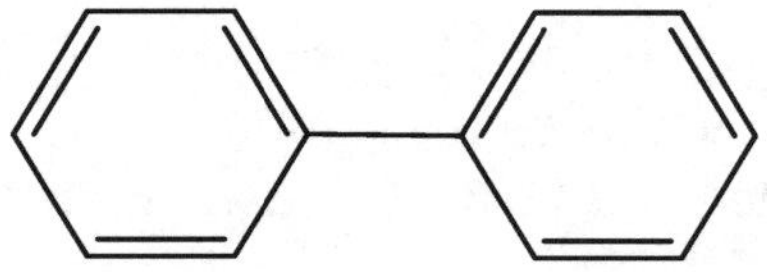

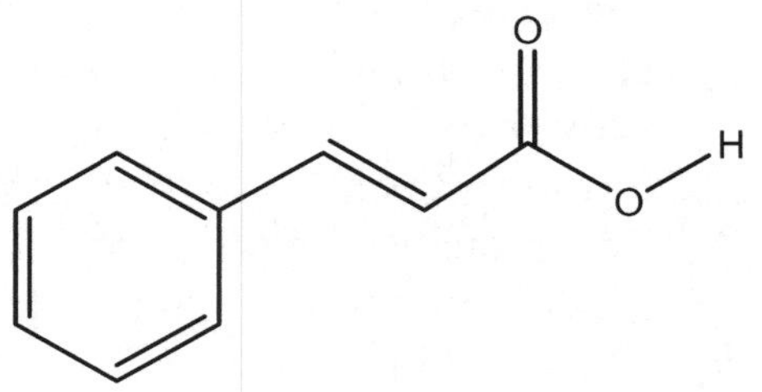

2. How could the purity of each compound be assessed?

Synthesis of n-butyl bromide an example of Nucleophilic Substitiution
Reaction: S_N2——EXP#7

Background
S_N2 reactions are different types of nucleophilic substitution reactions (S_N1). The general mechanism of nucleophilic substitution involves a species called a nucleophile (Nu) which donates an electron pair to the "substrate" carbon (usually an organic halide or tosylate) displacing a "leaving group" species (X). The S_N2 reaction is bimolecular where nucleophilic attack and loss of the leaving group occur simultaneously; the rate depends on both the concentration of substrate and of nucleophile.

Terminology
- **Substrate:** A molecule attacked by nucleophile (usually organic halides or tosylates)
- **Nucleophile:** An electron donor
- **Leaving group:** Species X: I,⁻ Br,⁻ Cl,⁻ TosO,⁻...etc.

General Mechanism

For S_N2: Priority decreases: $R^1 > R^2 > R^3$

Rate = K [Substrate] [Nucleophile]   Inversion
Nu: approaches from the opposite side of "X"

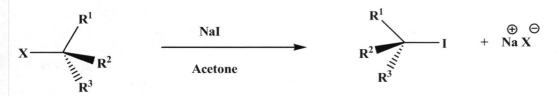

Factors Affecting the Rate of S_N2 Reaction
1. The structure of organic halides:
   Aryl halides < 3° << 2° < 1° < allyl halides ~ benzyl halides
   Reactivity increases--------------→
2. The structure of nucleophile: Larger nucleophiles are less effective (due to steric hindrance).
3. The effect of the solvent: The reactivity of the nucleophile affected by the solvent (in polar protic solvents a strong base is a weak nucleophile and in aprotic solvents, a good base is a good nucleophile).

- *In general, a good base is a good nucleophile, and a highly polarizable bases (I⁻, RS⁻ are better than a less polarizable bases (F⁻, Cl⁻, ⁻OCOR)*

Specific Reaction

$$\text{CH}_3\text{CH}_2\text{CH}_2\text{CH}_2\text{OH} \xrightarrow[\text{NaBr, Reflux}]{\text{H}_2\text{SO}_4} \text{CH}_3\text{CH}_2\text{CH}_2\text{CH}_2\text{Br}$$

Procedure

1. Weigh out a clean 5 mL conical vial and record it into your notebook.
2. Weigh out 1.1 g of NaBr using weighing paper and add it to the conical vial.
3. Add 1 mL of distilled water using a graduated cylinder.
4. Add 1 mL of 1-butanol using a graduated cylinder. After addition of the reagents, how many layers do you observe? Identify each layer.
5. Stir the solution using spatula until most of the salt dissolves. Also use spatula to break the clumpy salt formed after addition of water and butanol. Not all of the salt will dissolve. Mix the contents with a spatula for two minutes for a good yield.
6. After mixing, cool the contents in an ice water bath using 150 ml beaker. Make sure the level of the Ice-water in 150 ml beaker is lower than the height of the conical vial.
7. While in the bath and stirring the contents, add 1 mL of concentrated sulfuric acid **drop wise using a Pasteur pipette over a period of 5 minutes (save some (a few drops) for the next step)**.
8. Stir the contents with a spatula while the vial is still in the bath. This breaks up the salt. Then add the remaining sulfuric acid. Continue adding the remaining acid until the entire amount of acid is in the solution. Now add the spin vane and record the temperature of reaction vessel in your lab manual and also write dawn your observation.
9. Transfer the vial from the ice water bath into the central hole of the aluminum block. Then equip the Via lwith Hickman Still and water condenser as shown in demo. The aluminum block should be on The hot plate. Do not turn anything on yet. Simply set it up. The thermometer in the thermometer adapter should be long enough so that the mercury reservoir is just below the well of the Hickman Still.
10. Turn the water condenser on and start to distill the reaction mixture. Increase the temperature gradually. If the spin vane does not spin but instead vibrates this is still fine. **Record both the temperature Inside the apparatus and outside upon starting. Continue observing the temperature and recording both temperatures when the salts dissolve, when the solution boils, and when the vapor condenses into the well.**
11. Distill for one hour. Make sure to record the temperatures as mentioned previously. **If the temperature of the hot plate (external temperature) is higher than 180 °C then immediately turn the temperature setting on the hot plate down to 1.**
12. After distillation, turn off the hot plate but leave the distillation apparatus running. Allow the entire vapor to condense.
13. Collect the entire condensate in a pre-weighed 3 mL vial. Use a short stem Pasteur Pipette to draw out the condensate. If there is still some left then rinse the well with dichloromethane then draw all of it back out and into the vial.
14. After collecting the condensate in the vial wait for the dichloromethane to evaporate (air-dry) and then weigh the vial again.
15. There may be two layers visible in the solution. If so, add enough anhydrous sodium sulfate to Remove the top layer.
16. Perform a GC analysis by injecting 1 micro liter of the solution into the injector port of gas chromatograph.

17. Determine corresponding peaks for 1-butanol and 1-bromobutane (Refer to the handouts of GC traces of the starting material and product).
18. Calculate the percent composition of 1-Bromobutane in the solution from the following equation;

$$Percent\ Composition = \frac{Area\ of\ 1-Bromobutane}{Area\ of\ 1-Bromobutane\ +\ Area\ of\ 1-Butanol} \times 0.97 \times 100\%$$

Post-lab Questions

1. What would be the order of elution of your crude product mixture (a mixture of 1-bromobutane and 1-butanol), if you injected your distillate into the injector port of GC Shimadzu 2014 with RXI-1ms capillary column made of 100% dimethyl polysiloxane which is a "nonpolar column". Explain for full credit.
2. What would be the order of elution of your crude product mixture, if you injected your distillate into the injector port of GC Shimadzu 2014 Stabiwax Column a capillary column made of Carbowax polyethylene glycol which is a "polar column". Explain for full credit.
3. What are the formulas of the salts that precipitate when the reaction mixture was cooled? Give equations for their formations.
4. Why would it be undesirable to wash the halide product with aqueous sodium hydroxide?
5. What were the species in the distillate?
6. The distillate contained two layers. Identify each.

Elimination: E1 and E2 Reactions— Exp#8 and 9

Background
The E1 and E2 reactions are two types of elimination reactions. The general mechanism of elimination involves a "base" (B), which abstracts a proton from one carbon of a 'substrate' (usually an organic halide) leaving a lone pair of electrons. The electron pair then displaces a "leaving group" species (X), resulting in the formation of a new double bond. The difference between the E1 reaction and the E2 reaction is in the rate-determining step. The E1 reaction is a unimolecular process that proceeds through formation of an intermediate carbocation; the rate depends only on the concentration of substrate. The E2 reaction is bimolecular where deprotonation and loss of the leaving group occurs simultaneously; the rate depends on both concentration of the substrate and base.

Goal of the Experiment
1. Conduct a dehydration reaction (E1) of 2-butanol and determine the product ratio by GLC analysis.
2. Conduct a dehydrobromination reaction (E2) of 2-bromobutane and determine the product ratio by GLC analysis.

Terminology
- *Anti Periplanar*: A conformation in which an electron pair from a neighboring C-H bond pushes out the leaving group on the opposite side of the molecule (in a staggered conformation).
- *Zaitsev's rule*: A rule stating that E2 elimination reactions normally yield the more highly substituted alkene as major product.

General Reactions:
(a) E1 Reaction:

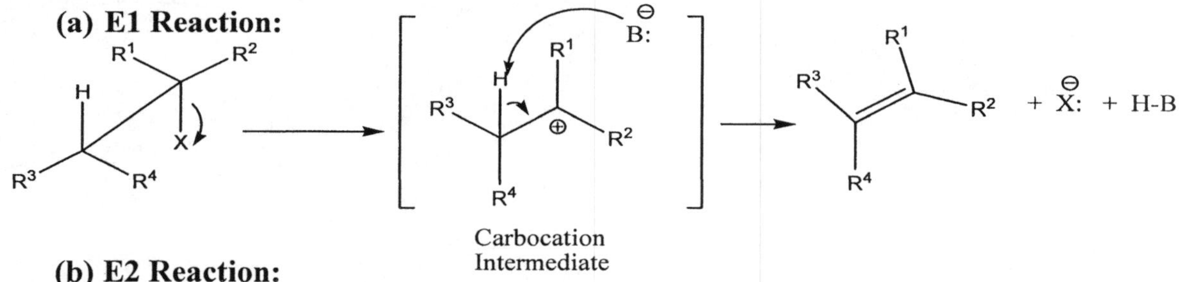

(b) E2 Reaction:

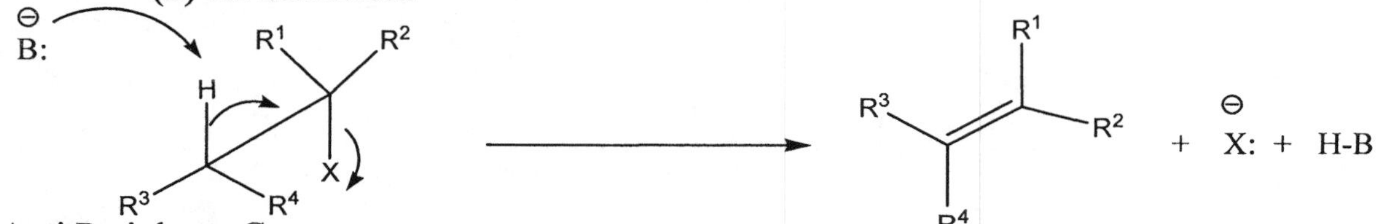

Anti Periplanar Geometry

✓ **NOTE**: The leaving group X and H must be in Anti Periplanar geometry.

Specific Reactions

(a) E1 Reaction:

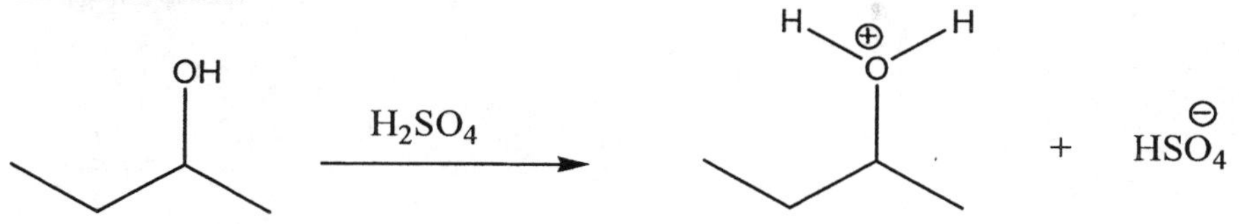

sec-butanol + H_2SO_4 → Oxonium ion + HSO_4^-

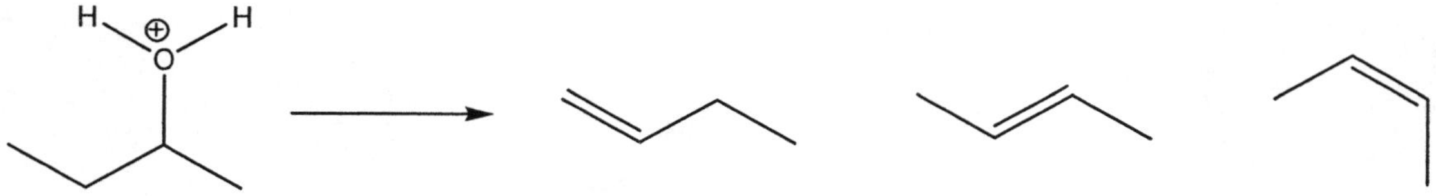

Oxonium ion → 1-butene + trans-2-butene + cis-2-butene

(b) E1 Mechanism:

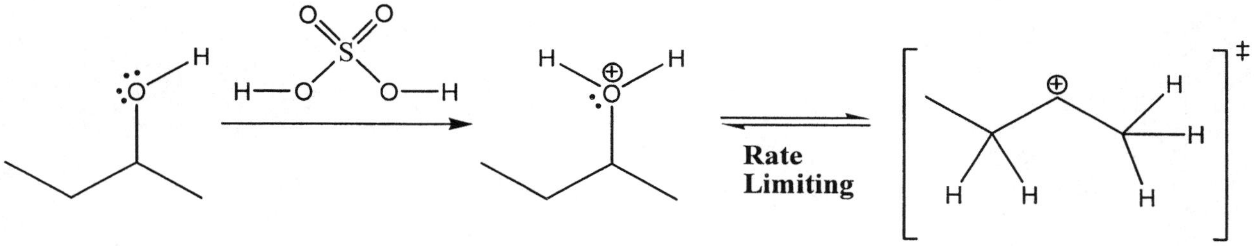

sec-butanol + H_2SO_4 → Oxonium ion ⇌ (Rate Limiting) [carbocation transition state]‡

It is possible for *EITHER* of these protons to be removed!

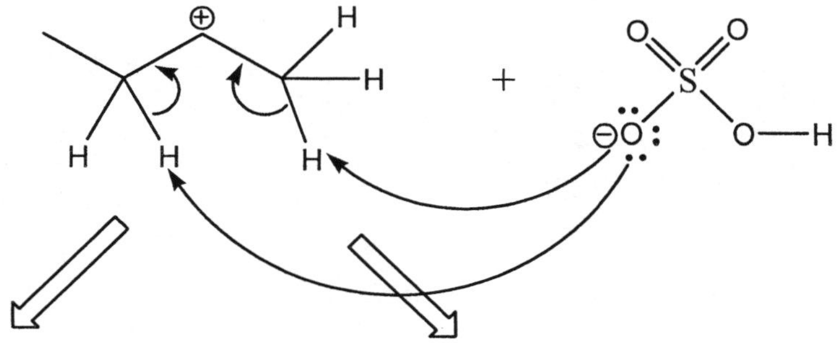

(trans) + (cis)
More substituted - Major products

Less substituted - Minor product

(c) E2 Reaction:

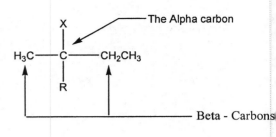

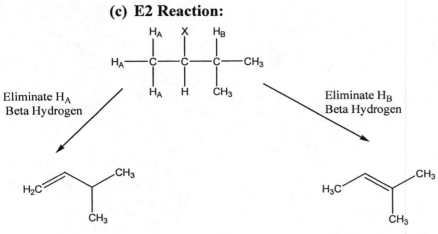

Zaitsev's product

(d) E2 Mechanism:

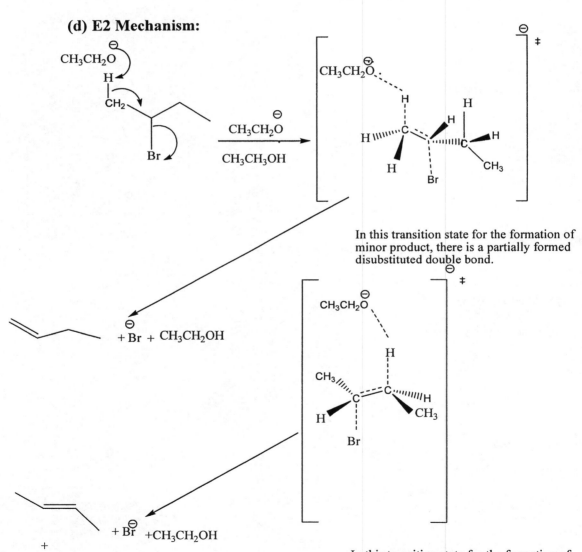

In this transition state for the formation of minor product, there is a partially formed disubstituted double bond.

In this transition state for the formation of major product, there is a partially formed disubstituted double bond.

Zaitsev's product

Factors That Affect Reaction Rate:
E1 Reaction
1. The structure of substrate: aryl-X<1° <2° <3 °< allyl-X< benzyl-X
   Reactivity increases ------------→
2. Leaving group reactivity: A general rule: A weak base is a better leaving group.
   $I^- > Br^- > H_2O = Cl^- >> F^-$
   Better Leaving Group   Worse Leaving group
3. The polarity of the solvent: A polar solvent favors the ionization step and stabilizes the ions formed.

E2 Reaction
1. The structure of substrate:
   Aryl halides <3° <2° <1° < allyl halides <benzyl halide
   Reactivity increases -------------------→
2. The stronger the base, the faster the reaction.
   In protic solvents, a strong base is highly solvated and stabilized, so it is not available to react. In aprotic solvents, a strong base is highly reactive.
3. The higher the temperature the faster the reaction.

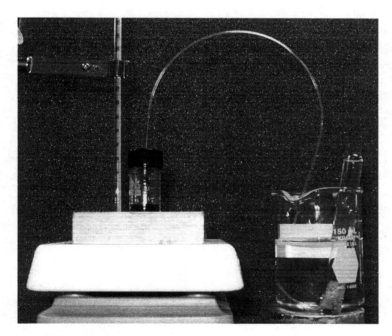

Figure 13: Experimental setup for E1 and E2 reaction

Procedure
E1 Reaction
1. Set up the reaction apparatus as demonstrated in Figure 13 and by your instructor-work in pairs. Your instructor must check your apparatus *before* you begin the experiment.
2. Place a spin vane in a 5 mL conical vial.
3. Using a pipette, add 3 drops of concentrated H_2SO_4 to the vial.
   ➤ **CAUTION:** ***CONCENTRATED H_2SO_4 IS CORROSIVE!!***

4. Add 24 drops of 2-butanol to the vial, cap the vial, and return to the apparatus set up at your bench.
5. While stirring, heat to 80°C and continue heating to 100 °C as necessary.
6. Once you have obtained a gas sample, using the Gowmac GLCs, run **ONE** GLC *only* (using 1 mL syringe, inject 1 mL of gas sample into the injector port of GC) and share the chromatogram with your lab partner.
7. Determine the area of the peak(s) in your chromatogram using the H x $W_{1/2}$ = area formula and determine the percent composition of each alkene product formed in the reaction.

E2 Reaction
1. Set up the reaction apparatus as demonstrated in Figure 13 and by your instructor-work in pairs. Your instructor must check your apparatus *before* you begin the experiment.
2. Place a spin vane in a 5 mL conical vial.
3. Using a pipette, add NaOEt solution up to the 1.0 mL mark on the vial.
    > **CAUTION:** *CARE SHOULD BE USED IN HANDLING ALL REAGENTS. THE ALKYL BROMIDE IS AN IRRITANT AND THE NaOEt SOLUTION IS CORROSIVE!!*
4. Add 10 drops of 2-bromobutane to the vial.
5. Cap the vial and immediately return to the apparatus at your bench.
6. The E2 elimination and the evolution of gas occur fairly rapidly.
7. Heat to 80 °C as necessary.
8. Once you have obtained the gas sample, using the Gowmac GLCs, run **ONE** GLC *only* (using 1 mL syringe, inject 1mL gas sample into the GC) and share the chromatogram with your lab partner. If you are using SRI 8610C, inject 10 µL gas sample into injector port of the GC.
9. Determine the area of the peak(s) in your chromatogram using the H x $W_{1/2}$ = area formula and determine the percent composition of each alkene product formed in the reaction.
    NOTE: Only one chromatogram per pair. The lab partner without the actual chromatogram should record the data in his/her lab notebook and show all the required calculations. Also, they should reference his/her partner in his/her lab notebook.

References
Organic Chemistry 7th ed.; John McMurry.
Introduction to Organic Laboratory Techniques: A Microscale Approach 3rd ed.; Pavia, Lampman, Kriz, and Engel.

Safety
1. Handle the concentrated *H_2SO_4* with care; it is **CORROSIVE!**
2. Handle the *NaOEt* with care, it is **CORROSIVE!**
3. Handle the *alkyl bromide* with care; it is lachrymator (an **EYE IRRITANT**)!
4. If exposure to **ANY** of the reagents occurs, be sure to wash hands and any other affected areas with plenty of soap and water.
5. Take precautions when handling **HOT** objects.

6. Turn **OFF** any hot plates that are no longer being used.
7. **Thoroughly clean (with soap for glassware) and rinse (final rinse of distilled water) all glassware and utensils used PRIOR to returning them to your lab drawer.**

Disposal
Rinse conical vial residue with 0.5 mL of acetone and dispose of the rinse into the bottle labeled as "Organic Liquid Wastes."
Return TEST TUBES, CORKS, and MICROTUBING APPARATUS to the front desk.

Postlab Questions
1. Discuss the outcome of the E1 reaction: Which alkene was formed as the major product? Why? Which was formed as the minor product? Why?
2. Discuss the outcome of the E2 reaction: Which was formed as the major product? Why? Which was formed as the minor product? Why? What percent of 1-butene product was formed? Would you expect this percent to be high or low? Why?

Addition Polymers—Exp#10

Background

A polymer is a giant molecule (macromolecule) made up of repeating structural units. The reaction that unites small molecules (called monomers) into a polymer is called polymerization. Among the natural polymers that are important to life processes are starch and cellulose (the monomer is glucose), proteins (the monomers are amino acids), and nucleic acids (the monomers are nucleotides). Synthetic polymers can be divided into two main classes, depending on their method of synthesis: *addition polymers*, or chain-growth polymers, and *condensation polymers*, or step-growth polymers.

Addition polymers are compounds that are synthesized by repetitious addition of a monomer to a growing polymer chain. This process can be accomplished by cationic, anionic, or free radical mechanisms. In this experiment you will synthesize polymers of styrene and methyl methacrylate by *free radical polymerization.*

Terminology

- **POLYMER**: made up of many small units called *monomers.*
- **BIOLOGICAL POLYMERS: important to life processes** (e.g., starch, cellulose, protein, nucleic acids).
- **SYNTHETIC POLYMERS: are man-made polymers** (e.g., nylon, polyesters, polyethylene, polypropylene, polystyrene, and polyvinyl chloride).
- **CONDENSATION POLYMERS: formed by the condensation of two bifunctional or polyfunctional monomers** through the elimination of some small molecules such as water or hydrogen chloride.
- **ADDITION POLYMERS: monomer is an alkene.** Double bonds open up and make carbon-carbon bonds to form a polymer, which is saturated (e.g., polypropylene, polystyrene).
- **INITIATOR:** A substance with an easily broken bond that is used to initiate a radical chain reaction.

General Mechanism (Addition Polymerization)

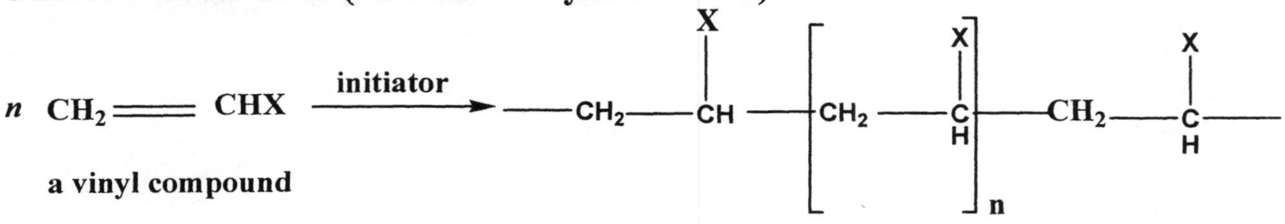

$$n \; CH_2 = CHX \xrightarrow{\text{initiator}} -CH_2-CH(X)-[CH_2-C(X)(H)-]_n-CH_2-C(X)(H)-$$

a vinyl compound → repeating units in a vinyl polymer, linked in a head-to-tail fashion

The reaction requires an unstable *initiator* (e.g., benzoyl peroxide or *t*-butyl benzoyl peroxide).

Application of Polymers in Industry

Monomer Name	Formula	Trade or Common Name of Polymer	Uses
Ethylene	$H_2C=CH_2$	Polyethylene	Packaging, bottles, cable insulation, films, and sheets
Propene (propylene)	$H_2C=CHCH_3$	Polypropylene	Automotive moldings, rope, carpet fibers
Chloroethylene (vinyl chloride)	$H_2C=CHCl$	Poly(vinyl chloride), Tedlar	Insulation, films, pipes
Styrene	$H_2C=CHC_6H_5$	Polystyrene, Styron	Foam and molded articles
Tetrafluoroethylene	$F_2C=CF_2$	Teflon	Valves and gaskets, coatings
Acrylonitrile	$H_2C=CHCN$	Orlon, Acrilan	Fibers
Methyl methacrylate	$H_2C=C(CH_3)CO_2CH_3$	Plexiglas, Lucite	Molded articles, paints
Vinyl acetate	$H_2C=CHOCOCH_3$	Poly(vinyl acetate)	Paints, adhesives

Polymerization via Free Radical

Step 1—Initiation (formation of the initiator): Initiation of the reaction takes place once several radicals are created via a catalyst. For example, when benzoyl peroxide is used as the initiator, the O-O bond is broken (via heat) to form two benzoyloxy radicals. A benzoyloxy radical then adds to the C=C bond of ethylene in order to generate a carbon radical. One of the electrons from the C=C double bond pairs with the odd electron located on the benzoyloxy radical to form a new C-O bond. The other electron remains on the carbon atom.

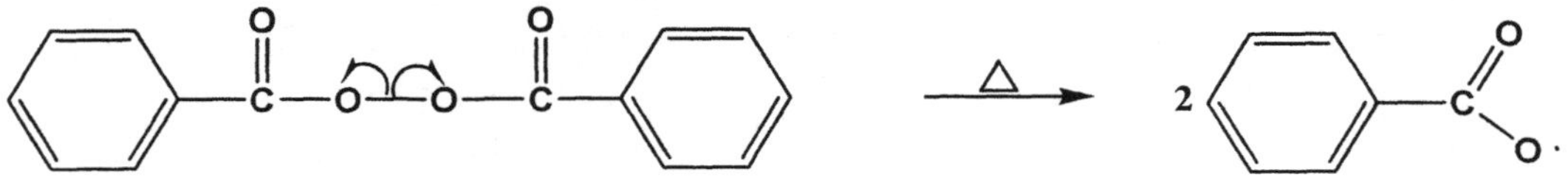

benzoyl peroxide $\xrightarrow{\Delta}$ 2 benzoyloxy radical = BzO•

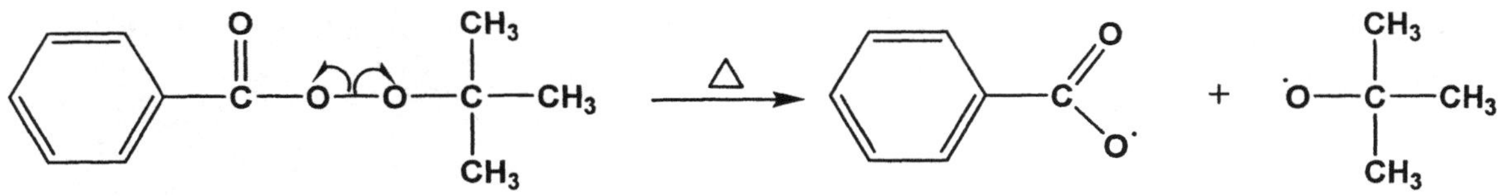

tertiary butyl benzoyl peroxide $\xrightarrow{\Delta}$ benzoyloxy radical = BzO• + •O−C(CH_3)_3

Step 2—Propagation:

BzO• + $H_2C=CH_2$ ⟶ BzO—CH_2CH_2•

Polymerization occurs when the carbon radical formed during initiation adds to another ethylene molecule—another radical is formed. The repetition of this reaction for hundreds or even thousands of times is what builds the polymer chain.

$$BzOCH_2\dot{C}H_2 + H_2C{=}CH_2 \longrightarrow BzOCH_2CH_2CH_2\dot{C}H_2 \xrightarrow{repeat} BzO(CH_2CH_2)_n CH_2\dot{C}H_2$$

Step 3—Termination: The polymerization chain process is ended by a reaction that consumes the radical instead of utilizing it to increase the polymer. There are two ways to terminate this polymerization reaction—*combination* and *disproportionation*.

For example, combination reaction:

$$2\ R{-}\dot{C}H_2CH_2 \longrightarrow R{-}CH_2CH_2CH_2CH_2{-}R$$

Specific Mechanism

Step 1: The radical *initiator* is generated from benzoyl peroxide using *heat* or *light*.

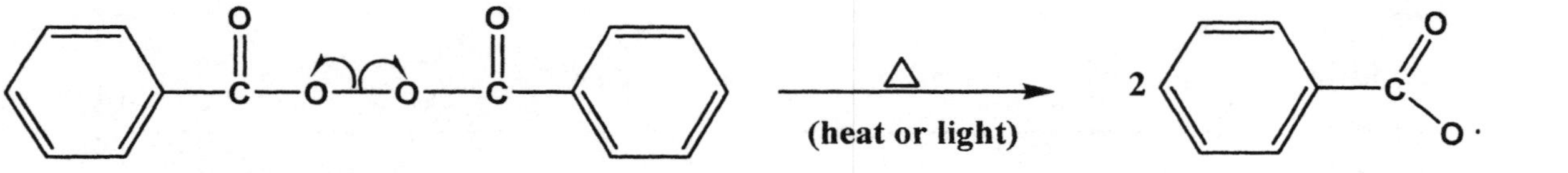

benzoyl peroxide → (heat or light) → 2 benzoyloxy radical = BzO•

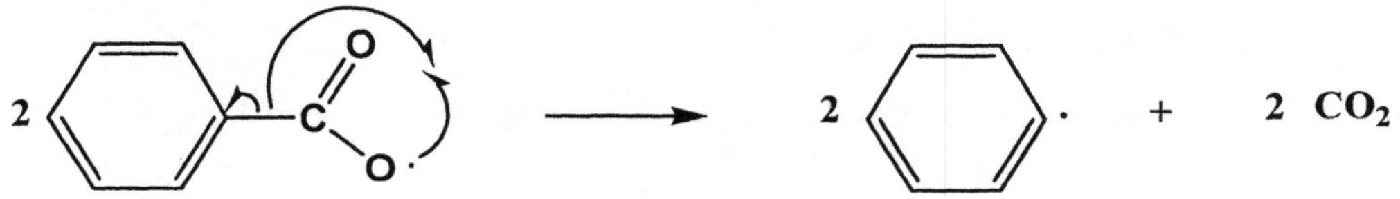

2 benzoyloxy radical = BzO• → 2 radical initiator + 2 CO$_2$

Step 2: Polymer *initiation* occurs by addition of the first monomer unit, methyl methacrylate. A new radical is formed in this step.

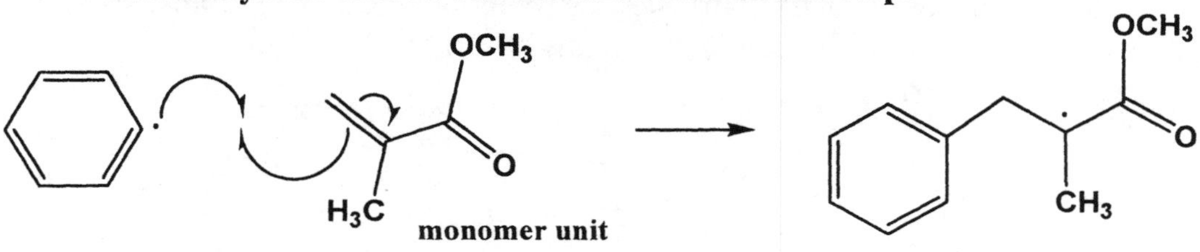

Step 3: *Propagation* of the polymer chain occurs by consecutive addition of monomer units.

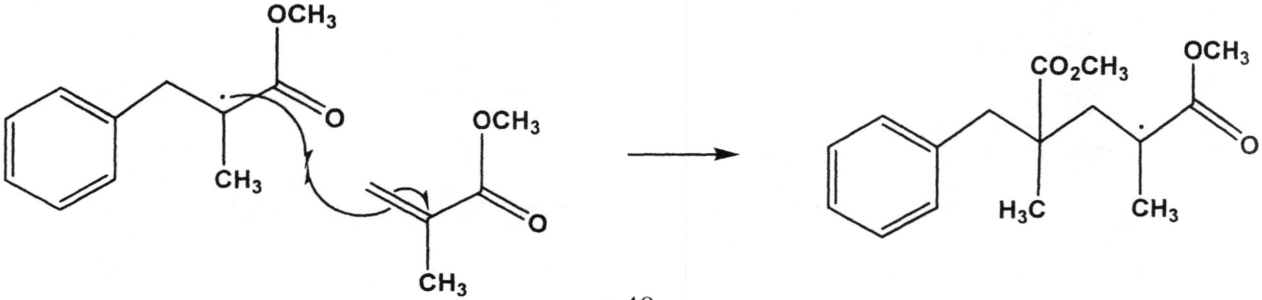

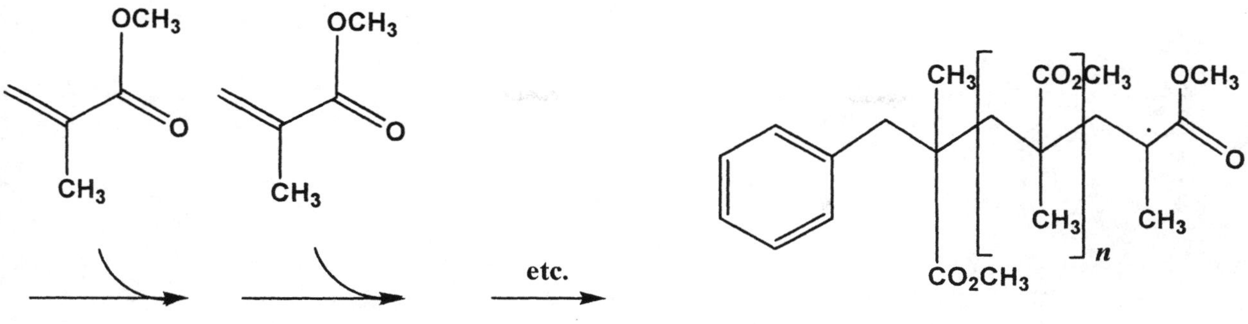

Step 4: *Termination* of the addition reaction can occur either by *disproportionation* of a radical polymer chain or by the *combination* of two chains.

DISPROPORTIONATION

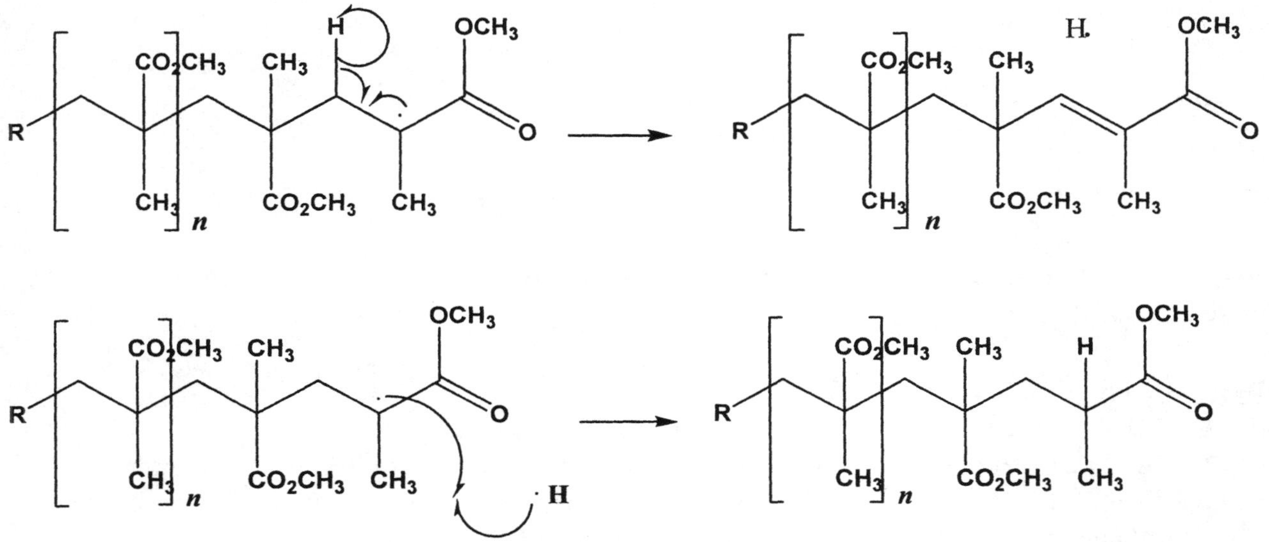

COMBINATION

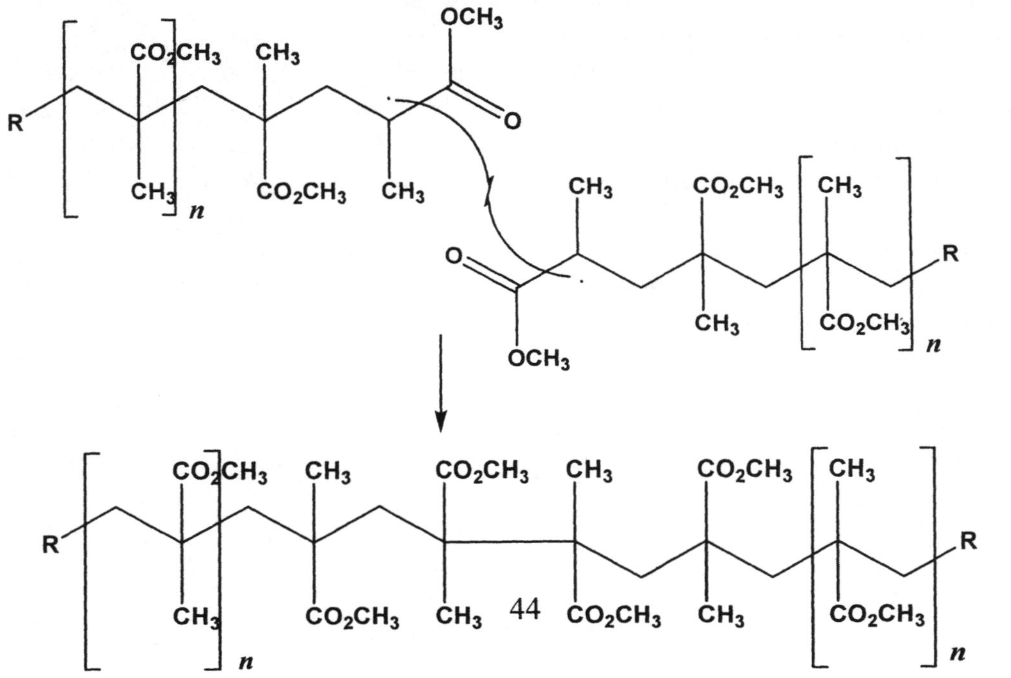

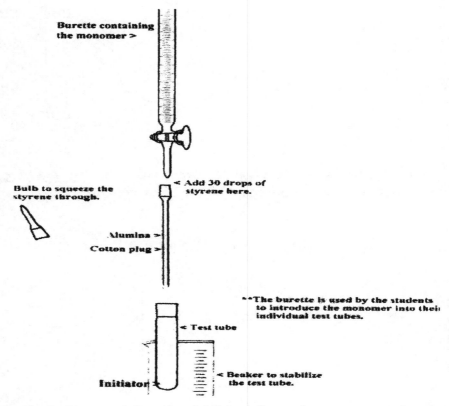

Figure 14: Procedure for removing the inhibitor from the monomer (styrene and methyl methacrylate).

Procedure

1. Fill a 50 mL beaker ~2/3 full with water and begin heating it.
2. The lab instructor will provide each student with a test tube with the initiator added by the instructor. A wooden spatula with a marking on it will be provided to you to measure out an approximate amount of initiator. *Refer to Figure 14.*
3. Obtain a Pasteur pipette and place a very small piece of cotton in it. The cotton is then pushed down with a copper wire to "plug up" the pipette. *The instructor may demonstrate this during the lecture part of lab.*
4. Alumina will be provided in each lab in large test tubes. You will need about 1 cm of alumina in your pipette to remove the inhibitor in the styrene. Hold the large test tube filled with alumina horizontal and slide the large open side of the pipette in. By following this procedure, you will obtain the necessary amount of alumina without having to try to spoon it in with a spatula.
5. Now, add 30 drops of styrene from the burette to your Pasteur pipette. Take a pipette bulb and gently squeeze the styrene liquid through the alumina and cotton directly into your test tube containing the initiator.
6. Swirl the test tube until all the initiator is dissolved, then cover the test tube with parafilm and poke a hole through the top.
7. The 50 mL beaker of water should be at a boil by now. Place the test tube in the hot water bath until polymerization occurs. Record the time for the polymerization process.

9. Repeat the procedure using methyl methacrylate and record the time for the polymerization process. Also compare the physical properties of each polymer with each other. In addition, explain the differences in time with respect to each polymer production.

References
Organic Chemistry 7th ed.; John McMurry.
Introduction to Organic Laboratory Techniques: A Microscale Approach 3rd ed.; Pavia, Lampman, Kriz, and Engel.

Safety
1. **Styrene vapor is very irritating to the eyes, mucous membranes, and upper respiratory tract. Do NOT breathe the vapors and AVOID getting this reagent on your skin. Exposure can cause nausea and headaches. All operations with styrene must be conducted in a hood.**
2. **Benzoyl peroxide is flammable and may detonate on impact or upon heating (or grinding). Please distribute using a wooden spatula. It should be weighed on glassine (glazed, not ordinary paper). Clean all spills with water.**
3. **Methylmethacrylate is highly flammable, a lachrymator, and irritating to the respiratory system and eyes. It may also cause sensitization upon contact with the skin.**
4. **Take precautions when handling HOT objects.**
5. **If exposure to ANY of the reagents occurs, be sure to wash hands and any other affected areas thoroughly with plenty of soap and water.**
6. **Turn OFF any hot plates that are no longer being used.**

Disposal
Dispose of the test tubes containing polymer in broken glass container. (**NOTE**: Before disposing of the test tubes, make sure they do not contain any leftover monomer; consult your instructor before disposing of the contents.) Next, thoroughly clean (with soap for glassware) and rinse (final rinse of distilled water) ALL glassware and utensils used PRIOR to returning them to your lab drawer.
Dispose of the disposable test tubes and microfiltration columns in the broken glass container.
Dispose of the used parafilm in the trash.

Postlab Questions
1. Why does the mixture of styrene and initiator become viscous (thicker) and eventually solidify as the polymerization reaction proceeds? Explain in terms of intermolecular interaction.
2. Using correct and complete arrow formalism, write out the complete mechanism for free radical polymerization of styrene.
3. Explain why head-to-tail addition is energetically more favorable than head-to-head addition?
4. In which step of the free radical polymerization might head-to-head units be incorporated into the polymer chain?

Diels-Alder Reaction—Exp#11

Background
The *Diels-Alder* reaction is a very useful reaction in organic chemistry. It is also called a *[4+2]π cycloaddition reaction*. This means a conjugated diene, containing 4π electrons, is combined with another alkene, containing 2π electrons, to form a ring. The 4 electron component is referred to as the **diene** and the 2 electron component is called **dienophile**. The diene adds to the dienophile by the *concerted* mechanism. In other words, all new bonds are formed simultaneously.

Goal of the Experiment
To synthesize cis-4-cyclohexene-1,2-dicarboxylic anhydride using [4+2]π cycloaddition reaction (Diels-Alder reaction).

General Reaction and Mechanism

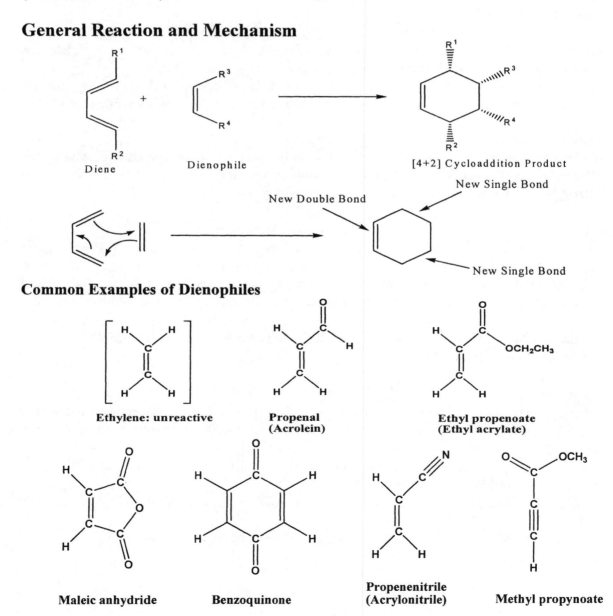

Specific Reaction and Mechanism

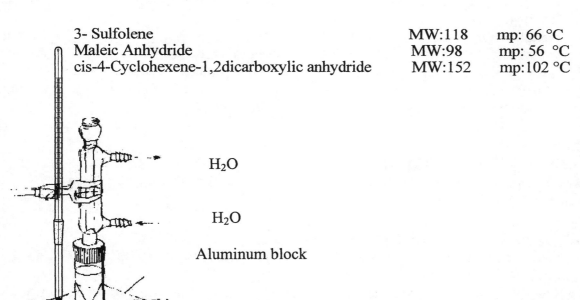

3- Sulfolene	MW:118	mp: 66 °C
Maleic Anhydride	MW:98	mp: 56 °C
cis-4-Cyclohexene-1,2dicarboxylic anhydride	MW:152	mp:102 °C

Figure 15: Diels-Alder setup

Experimental Procedure
1. Weigh out 1.0 mmol of maleic anhydride and 2.2 mmol of 3-sulfolene and transfer each solid to a clean, dry 3 mL conical vial. Record the exact amount of each solid weighed (in mg and mmol) in the notebook.

2. To the vial, add xylene up to the 0.3 mL mark by pipette.
3. Add one Teflon boiling chip to the vial and attach a condenser as displayed in Figure 15.
4. Begin heating the mixture at a steady rate (approximately 5 degrees per minute) until the temperature reaches 180 °C.
5. As you are heating, make sure that all the reagents dissolve by shaking or swirling the vial. When the temperature reaches 180 °C, turn the heat down slightly and monitor the time so that the reaction is stirred between 180 °C and 190 °C for 20 minutes.
✓ *CAUTION: YOU MUST MONITOR THE TEMPERATURE CAREFULLY SO THAT THE REACTION IS NOT OVERHEATED.*
   NOTE: If your crude product is colorless and clear, skip step 6 and 7.
6. Once the heating period is over, allow the vial to slowly cool to room temperature. Meanwhile, prepare a micro filtration column by inserting a small cotton plug into a pipette and packing it firmly with a piece of copper wire.
7. To the pipette, add 5 mm of sand, 10 mm of Celite, 5 mm of charcoal, and another 2 mm of sand. Clamp the micro column to a ring stand over a Craig tube.
8. To the cooled reaction mixture (3 mL vial), add 20–25 drops of toluene and transfer the entire mixture to the micro column.
9. Collect the filtrate in the Craig tube.
10. Rinse the 3 mL vial with a few more drops of toluene and pass the rinse solution through the micro column and into the Craig tube.
11. To the filtrate (Craig tube), add hexane dropwise while swirling the tube, until the product begins to precipitate.
12. Recrystallize the product by carefully heating the solution in the Craig tube in a warm water bath and then allowing it to slowly cool to room temperature.
13. Finally, cool the solution in an ice bath and collect the product by centrifugation.
14. Air-dry the crystals for 10 minutes and determine the melting point, theoretical yield, actual yield, and percent yield of the Diels-Alder product.

References
Organic Chemistry 7th ed.; John McMurry.
Introduction to Organic Laboratory Techniques: A Microscale Approach 3rd ed.; Pavia, Lampman, Kriz, and Engel.

Safety
1. **Be sure to cover the centrifuge top while using it.**
2. **Be sure to balance the centrifuge with either another group's sample or with a tube containing an equivalent amount of water.**
3. **Thoroughly clean (with soap for glassware) and rinse (final rinse of distilled water) all glassware and utensils used PRIOR to returning them to your lab drawer.**

Disposal
Dispose of any excess reagent and/or product in the "Organic Liquid Waste" receptacle.
Dispose of Pasteur pipettes in the glass disposal bin.

Postlab Questions
1. Draw the mechanism for the conversion of 3-sulfolene to 1,3-butadiene.
2. What reagents (diene and dienophile) would you need to prepare 1-methyl-4-propylcyclohexene?

Competitive Aromatic Nitration—Exp#12

Background
Aromatic nitration is one example of *electrophilic aromatic substitution.* An "electrophile" is a species that accepts an electron pair from another atom. Electrophilic substitution means that an electrophile, rather than a nucleophile, replaces another atom in a molecule. The nitration of benzene, depicted below, serves to illustrate one possible mechanism for electrophile aromatic substitution. The electrophile in the reaction is called a nitronium ion, which is formed by combining nitric acid and sulfuric acid.

The Goal of the Experiment
To conduct a competitive aromatic nitration reaction, to compare the reactivities of methyl benzoate and acetanilide towards nitration by measuring the m.p. of the product, and to compare this experimental value to the literature value for the expected products.

Reactivity and Orientation
The reactivity of the aromatic ring towards electrophilic substitution depends upon the tendency of a substituent group to release, or withdraw electron density. Groups that donate electron density into the ring (e.g., –OCH$_3$) will activate the *ortho* and *para* positions on the ring, towards addition of the electrophile. Groups that withdraw electron density (e.g., –CO$_2$R) will activate the meta positions on the ring. Note that substituents with an electron-withdrawing resonance effect have the general structure –Y=Z, where the Z atom is more electronegative than Y.

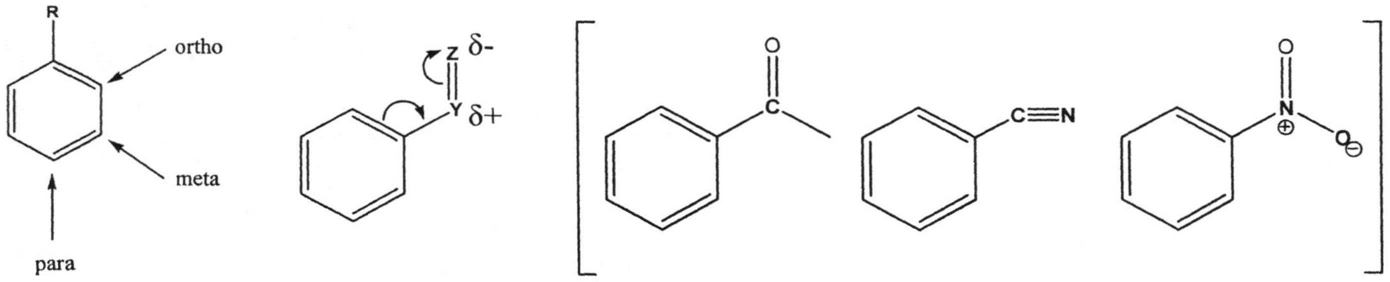

Substituents with an electron-donating resonance effect have the general structure —Y, where the Y atom has a lone pair of electrons available for donation to the ring.

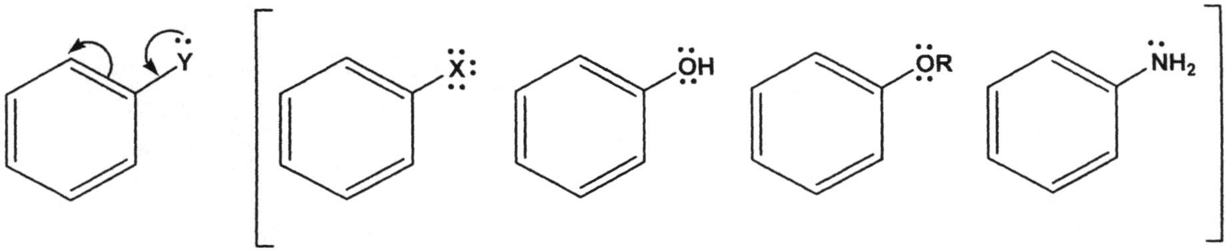

Substituent Effects in Electrophilic Aromatic Substitution

Substituent	Reactivity	Orientation	Inductive effect	Resonance effect
—CH$_3$	Activating	Ortho, para	Weak; electron-donating	None
—ÖH, —N̈H$_2$	Activating	Ortho, para	Weak; electron-withdrawing	Strong; electron-donating
—F̈:, —C̈l:, —B̈r:, —Ï:	Deactivating	Ortho, para	Strong; electron-withdrawing	Weak; electron-donating
—N⁺(CH$_3$)$_3$	Deactivating	Meta	Strong; electron-withdrawing	None
—NO$_2$, —CN, —CHO, —CO$_2$CH$_3$, —COCH$_3$, —CO$_2$H	Deactivating	Meta	Strong; electron-withdrawing	Strong; electron-withdrawing

General Reaction

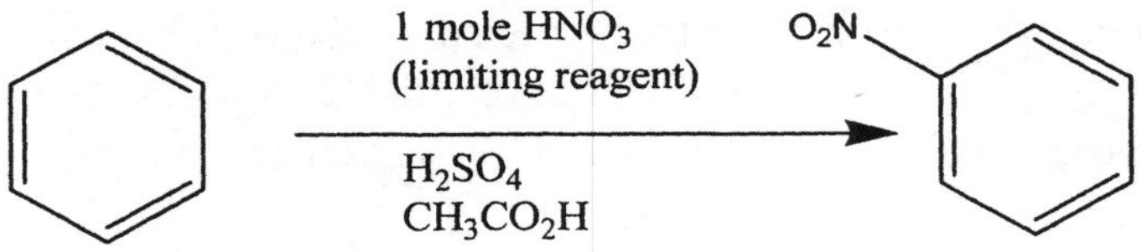

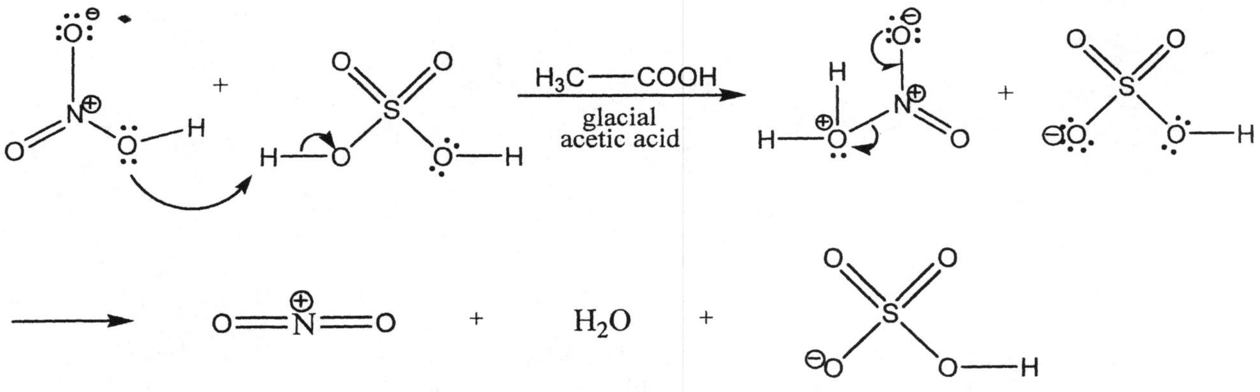

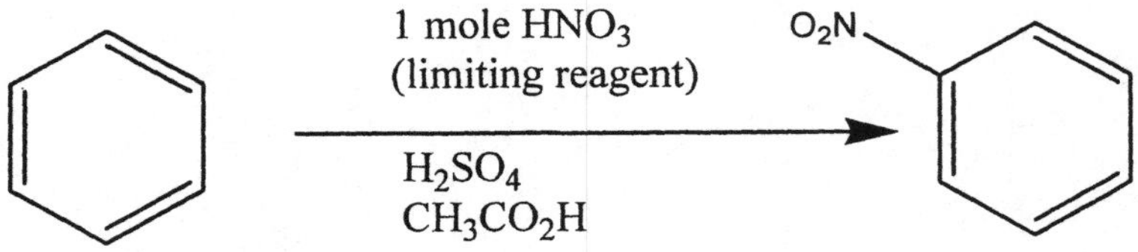

$$C_6H_6 \xrightarrow[\text{CH}_3\text{CO}_2\text{H}]{\substack{\text{1 mole HNO}_3 \\ \text{(limiting reagent)} \\ \text{H}_2\text{SO}_4}} C_6H_5NO_2$$

[Mechanism: HNO$_3$ + H$_2$SO$_4$ with glacial acetic acid → protonated HNO$_3$ + HSO$_4^-$]

$$\longrightarrow \quad O={\overset{\oplus}{N}}=O \quad + \quad H_2O \quad + \quad HSO_4^-$$

$$C_6H_6 \xrightarrow[\text{CH}_3\text{CO}_2\text{H}]{\substack{\text{1 mole HNO}_3 \\ \text{(limiting reagent)} \\ \text{H}_2\text{SO}_4}} C_6H_5NO_2$$

$$\text{HONO}_2 \; + \; 2\,\text{H}_2\text{SO}_4 \; \rightleftharpoons \; \text{H}_3\overset{\oplus}{\text{O}} \; + \; \overset{\oplus}{\text{NO}}_2 \; + \; 2\,\text{HSO}_4^{\ominus}$$
nitronium ion

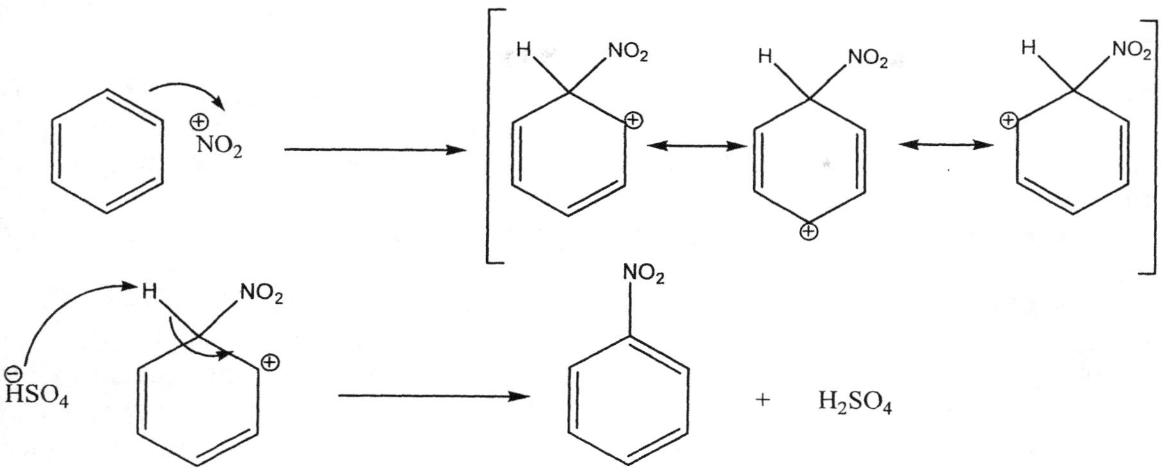

Specific Reaction and Mechanism

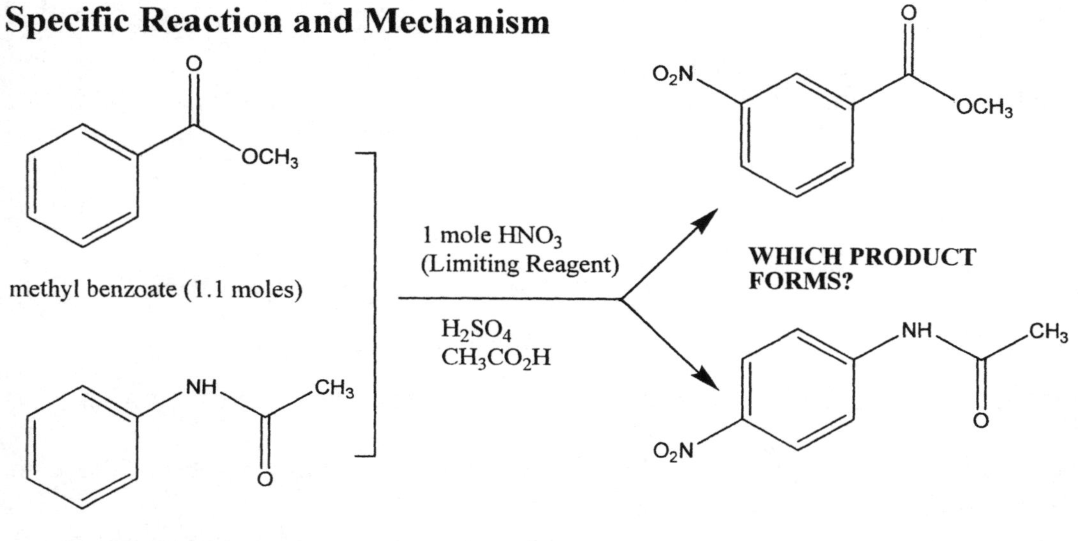

methyl benzoate (1.1 moles)

acetanilide (1.1 moles)

1 mole HNO₃ (Limiting Reagent)

H₂SO₄
CH₃CO₂H

WHICH PRODUCT FORMS?

Experimental Procedure

1. Weigh your 3 mL conical vial, with the screw cap on and record the mass on the worksheet.
2. In the hood, add 2 drops of concentrated nitric acid (HNO₃) to a 3 mL vial and put the screw cap back on.
3. Reweigh the vial and record the new mass on the worksheet.
4. Carry out the calculations as indicated on the worksheet to determine the amounts of methyl benzoate and acetanilide required.
5. In the hood, add 10 drops of glacial acetic acid and 20 drops of concentrated H₂SO₄ to the HNO₃ contained in the 3 mL vial and set aside.
6. Place a 5 mL conical vial with a spin vane onto the balance and weigh out the calculated amount of methyl benzoate. Since methyl benzoate is a liquid, you must carefully add it drop wise to your 5 mL vial in order to measure it accurately.
7. On a piece of weighing paper, accurately weigh out the calculated amount of acetanilide (solid) and transfer it to a 5 mL vial.
8. In the hood, add 10 drops of glacial acetic acid and 20 drops of H₂SO₄ to a 5 mL vial.
9. Place the 5 mL vial into a 100 mL beaker containing 25 mL of ice water on a stir-plate and begin stirring.

10. Using a pipette, carefully and slowly add the nitrating agent (contained in the 3 mL vial) to the 5 mL vial dropwise, over a period of 3–4 minutes.
11. Remove the 5 mL vial from the ice bath and stir at room temperature for an additional 10 minutes. Continue stirring and add a few small pieces of ice to the vial.
12. Once the white solid has precipitated, filter it using a Hirsch funnel and a side-arm test tube.
13. Rinse the solid 5–6 times with cold water to remove any excess methyl benzoate. Allow the vacuum to run for a few more minutes in order to completely dry the solid product.
14. Transfer the product to a 10 mL Erlenmeyer flask and add a minimum amount of hot ethanol to dissolve. This may take some time and the ethanol definitely needs to be hot!
15. Once all solid has dissolved, add 1–2 drops of water until a slight precipitate forms, then heat the mixture again to dissolve. This technique may help the recrystallization.
16. Cool the flask to room temperature, then in an icebath.
17. Once the product has recrystallized, filter it again using a Hirsch funnel. Collect the crystals and determine the melting point range, identity, theoretical yield, actual yield, and percent yield of the nitrated product. (Complete the work sheet.)

References
Organic Chemistry 7th ed.; John McMurry.
Introduction to Organic Laboratory Techniques: A Microscale Approach 3rd ed.; Pavia, Lampman, Kriz, and Engel.

Safety
1. **Nitric acid and sulfuric acid are very <u>CORROSIVE</u> substances, especially when mixed.**
2. **Take precautions to <u>NOT</u> get these acids on your skin. If you do get some of these on your skin, flush the affected area liberally with water.**
3. **Rinse all glassware, including the Hirsch funnel, with ~0.5 mL of acetone.**
4. **Following the acetone rinse, thoroughly clean (with soap for glassware) and rinse (final rinse of distilled water) <u>ALL</u> glassware and utensils <u>PRIOR</u> to returning them to your lab drawer.**
5. **Be sure to use wet paper towels or sponges to remove any traces of acids/reagents from the work benches.**

Disposal
Dispose of any excess reagents in the "Organic Liquid Waste" receptacle.
Dispose of the solid crystal product in the "Organic Solid Waste" receptacle.
Dispose of pipettes and capillary tubes in the broken glass container.

Postlab Questions
1. Explain why the –NH-CO-CH$_3$ group is an ortho/para director while the –COOCH$_3$ group is a meta director using carbocation intermediates formed after electrophilic addition.
2. Which product do you expect to form in the nitration reaction: methyl meta-nitrobenzoate or para-nitro acetanilide? Why?
3. Name the possible side products of this reaction and explain how you would remove them from your desired products.

** **NOTE**: You must include **COMPLETE** answers for the Postlab Questions in your lab notebook and also include the **COMPLETED** data and results sheet when you turn in your lab report. General and specific mechanisms/reactions **DO NOT** need to be included in your lab report. **DO NOT** include written procedure too.

Data Sheet and Results for Competitive Nitration Reaction

Limiting reagent: MW of acetanilide:

MW of methyl benzoate: MW of nitric acid:

1. Mass of empty 3 ml vial and screw cap: _____

2. Number of drops of Conc. HNO_3:   _____

3. Mass of 3mL vial and screw cap + Conc. HNO_3:   _____

4. Mass of Conc. HNO_3:   _____

5. Mass of Conc. HNO_3 in mg: _____

6. Actual mass of HNO_3 (NOTE: Conc. HNO_3 is 70%): _____

7. Mmoles of HNO_3:   _____

8. Mmoles of acetanilide: (7) x 1.1 equiv: _____

9. Mass of acetanilide required in mg: _____

10. Mmoles of methyl benzoate required: (7) x 1.1 equiv: _____

11. Mass of methyl benzoate required in mg: _____

12. MW of 4-nitro acetanilide: _____mp:____

13. MW of methyl 3-nitrobenzoate: _____mp:____

14. Melting point range of product: _____

15. Identity of the product: _____

16. Theoretical yield: _____

17. Actual yield: _____

18. Percent yield: _____

Based on your results, which reagent is more reactive towards electrophilic aromatic nitration?

Solubility and Solution——Exp#13

Background
The solubility of a solute in a solvent is the most important chemical concept underlying three basic techniques you have studied in the Organic Chemistry Laboratory (CHE 241L): crystallization, extraction, and chromatography. Also the solubility behavior will help you understand what is going on during synthesis, especially when there is more than one liquid phase present or when a precipitate is formed.

Goal of the experiment
The major goal of this experiment is to explain how to predict whether a substance will be soluble in a given solvent. This is not always easy, even for an experienced chemist. However, guidelines will help you make a good guess about the solubility of a compound in a specific solvent. In discussing these guidelines, it is helpful to divide the types of solution into two categories: molecular and ionic solutions.

A. Solutions in Which the Solvent and Solute Are Molecular
A useful generalization in predicting solubility is the widely used rule "Like dissolves like." According to this rule, a polar solvent will dissolve polar (or ionic) compounds, and a non-polar solvent will dissolve non-polar compounds. The reason for this behavior involves the nature of intermolecular forces of attractions such as dipole-dipole interaction, van der Waals forces (London or dispersion forces). Consult your lecture textbook for more information on these forces. To apply the rule "Like dissolves like," you must first determine whether a substance is polar or non-polar. The polarity of a compound is dependent on both the polarities of the individual bonds and the shape of the molecule. For most organic compounds, evaluating these factors can become quite complicated because of the complexities of the molecules. However, it is possible to make some reasonable predictions just by looking at the types of atoms that a compound possesses. As you read the following guidelines you will learn that polarity is a relative matter ranging from non-polar to highly polar.

How to determine the degree of solubility for organic compounds

Consider the solubilities for the following organic compounds at room temperature in water:

Cholesterol   0.002 mg / mL
Caffeine   22 mg / mL
Citric acid   620 mg / mL

In a typical test for solubility, 40 mg of solute is added to 1 mL of solvent. If all the material dissolves in the solvent by stirring for one minute, it is called soluble. Therefore, if you were testing the solubility of these three compounds, cholesterol would be insoluble, caffeine would be partially soluble and citric acid would be soluble.

B. Solutions in Which the Solute Ionizes and Dissociates
Many ionic compounds are highly soluble in water because of the strong attraction between ions and the highly polar water molecules. This also applies to organic

compounds that can exist as ions. For example, sodium acetate consist of Na^+ and CH_3COO^- ions, which are highly soluble in water. You may assume (some exceptions) that all organic compounds that are in ionic forms will be water soluble.

The most common way by which organic compounds become ions is in acid –base reactions. For example, carboxylic acids can be converted to water–soluble salts when they react with dilute aqueous NaOH:

$$CH_3(CH_2)_5COOH + NaOH \longrightarrow CH_3(CH_2)_5COO^-Na^+ + H_2O$$

   Water-insoluble carboxylic acid   Water-soluble salt

Amines, which are organic bases, can also be converted to water-soluble salts when they react with dilute aqueous HCl.

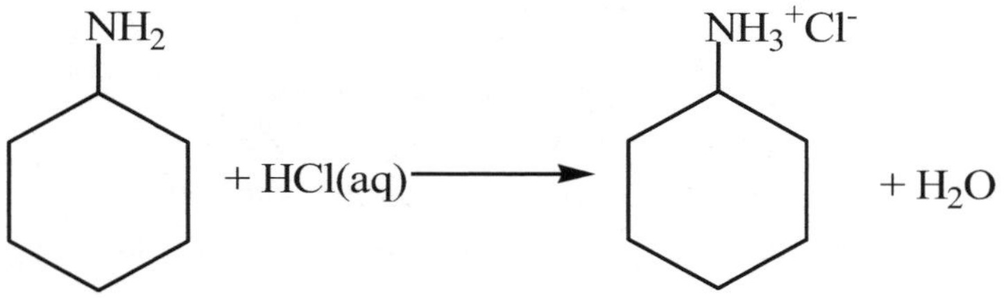

$$\text{C}_6\text{H}_{11}\text{NH}_2 + HCl(aq) \longrightarrow \text{C}_6\text{H}_{11}\text{NH}_3^+Cl^- + H_2O$$

   Water-insoluble amine   Water-soluble salt

Guidelines for Predicting Polarity and Solubility

1. All hydrocarbons are nonpolar.
   Examples: Hexane and Benzene
   Hydrocarbons such as benzene are slightly more polar than hexane because of their pi bonds.

2. Compounds possessing the electronegative elements oxygen or nitrogen are polar. Examples:

   $CH_3CH_2NH_2$ $CH_3CH_2OCH_2CH_3$ CH_3CH_2OH CH_3COCH_3

   Ethylamine   Diethyl ether   Ethyl alcohol   Acetone

   $CH_3COOCH_2CH_3$

   Ethyl acetate

   The polarity of these compounds depends on the presence of polar bonds such as C—O, C=O, OH, NH, and CN. The compounds that are most polar have NH or OH bonds. Therefore, they are capable of forming hydrogen bonds. Consult your textbook and TA regarding this concept.

3. The presence of halogen atoms, even though their electronegativities are relatively high, does not alter the polarity of an organic compound in a significant way. Therefore, these compounds are only slightly polar.
   The polarities of these compounds are more similar to those of hydrocarbons, which are nonpolar, than to that of water, which is highly polar.
   Examples:

   CH_2Cl_2

   Methylene chloride (dichloromethane)

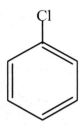

Chlorobenzene

4. When comparing organic compounds within the same family, note that adding carbon atoms to the chain decreases the polarity. For example, methyl alcohol (CH$_3$OH) is more polar than propyl alcohol (CH$_3$CH$_2$CH$_2$OH). Why?
5. Compounds having four or fewer carbons and containing oxygen or nitrogen are often soluble in water. Compounds containing five or six carbons and containing one of these elements are often insoluble in water or have borderline solubility.
6. Another factor that can affect solubility is the degree of branching of the alkyl chain in a compound. Branching of the alkyl chain in a compound lowers the inter- molecular forces between the molecules. This is usually reflected in a greater solubility in water for the branched compound than for the corresponding straight –chain compound.
7. The solubility rule ("Like dissolves like") may be applied to organic compounds that belong to the same family. However, this generalization may not apply if there is a substantial difference in size between the two compounds. For example, cholesterol, an alcohol with a molecular weight (MW) of 386.64, is only slightly soluble in ethanol where as 1-octanol is soluble in ethanol.
8. The higher the melting point is, the less soluble the compound. For instance, p-nitrobenzoic acid (mp 242 °C) is, by a factor of 10, less soluble in a fixed amount of ethanol than the ortho (mp 147 °C) and meta (mp 141- °C) isomers.

You can check your understanding of some of these guidelines by studying the list given below.

Compounds in increasing order of polarity

Increasing Polarity

Aliphatic hydrocarbons
  Hexane (nonpolar)
Aromatic hydrocarbons (π bonds)
  Benzene (nonpolar)
Halocarbons
  Methylene chloride (slightly polar)
Compounds with polar bonds
  Diethyl ether (slightly polar)
  Ethyl acetate (intermediate polarity)
  Acetone (intermediate polarity)
Compounds with polar bonds and hydrogen bonding
  Ethyl alcohol (intermediate polarity)
  Methyl alcohol (intermediate polarity)
  Water (highly polar)

Solvents in increasing order of polarity

Increasing polarity (Approximate)

RH	Alkanes (hexane, petroleum ether)
ArH	Aromatics (benzene, toluene)
ROR	Ethers (diethyl ether)
RX	Halides (CH_2Cl_2 > $CHCl_3$ > CCl_4)
RCOOR	Ester (ethyl acetate)
RCOR	Aldehydes, ketones (acetone)
RNH_2	Amines (triethylamine, pyridine)
ROH	Alcohols (methanol, ethanol)
$RCONH_2$	Amides (N,N-dimethylformamide)
RCOOH	Organic acids (acetic acid)
H_2O	Water

Terminology

- Soluble: It describes the solubility behavior of a solid substance in a solvent (dissolved).
- Insoluble: Not dissolved
- Miscible: It describes the solubility of a liquid solute in a solvent (dissolved).
- Immiscible: It describes the solubility of a liquid solute in a solvent (not dissolved).
- Solubility: It may be expressed in terms of grams/milligrams of solute per liter /- milliliter of solvent.

Experimental Procedure (Part A. Solubility of Solid Compounds)

1. Place about 40 mg (0.040 ±0.002 g) of grounded (powder) benzophenone into each of four dry test tubes.
2. Label the test tubes and then add 1 mL of water to the first tube, 1 mL of methyl alcohol to the second tube, and 1 mL of hexane to the third tube. The fourth tube will serve as a control.
3. Determine the solubility of each sample whether is partially soluble, soluble or insoluble by following the instruction of your TA. You should compare each tube with the control in making these determinations.
   Record these results in your notebook in the form of a table, as shown on the Next page.
4. Now repeat the above directions, substituting malonic acid and biphenyl for benzophenone. Record your results in your notebook.
   Note: you need to draw **the structure of each solute** in your lab notebook for full credit.

Organic Compounds	Water (highly polar)	Methyl Alcohol (Intermediate polarity)	Hexane (non-polar)	Prediction/ Rationale	Observation/ Conclusion
Benzophenone					
Malonic acid					
Biphenyl					

Note: Consistent stirring should be done for each solubility test.

Experimental Procedure (Part B. Solubility of Different Alcohols)

1. Take two test tubes and then add 1 mL of water to the first tube, 1 ml of hexane to the second test tube.
2. Then add one of the alcohols, dropwise to the first test tube and carefully observe what happens as you add each drop. Record your observations as directed by your TA in your lab notebook. Continue adding alcohol with shaking until you have added a total of 20 drops.
3. Repeat the procedure add the same alcohol to the second test tube containing hexane.
4. Repeat the above procedure using other alcohols listed in the following Table.
5. Record your results as soluble, insoluble, and partially soluble in your lab notebook.

Solvents

Alcohols	Water	Hexane	Predictions /Rationale	Observations / Conclusion
1-Octanol				
1-Butanol				
Methyl alcohol				

Experimental procedure (Part C. Miscible or Immiscible pairs)
1. Take 5 test tubes and label each one.
2. Add 1mL of each of the following pairs of liquids to each test tube.
3. Shake each test tube for 10-20 seconds to determine whether the two liquids are miscible (form one layer) or immiscible (form two layers).
4. Record your results in your notebook.
   Water and ethyl alcohol
   Water and diethyl ether
   Water and methylene chloride
   Water and hexane
   Hexane and methylene chloride

Experimental procedure (Part D. Solubility of Organic Acids and Bases)
1. Take three test tubes and label each.
2. Place about 30 mg (0.030 g) of benzoic acid into each of three dry test tubes.
3. Add 1 mL of water to the first tube, 1 mL of 1.0 M NaOH to the second tube, and 1 ml of 1.0 M HCl to the third tube. Stir the mixture in each test tube with a microspatula for 10-20 seconds.
4. Note whether the compound is soluble or is insoluble. Record your results in a table form.
5. Take the tube containing benzoic acid and 1.0 M NaOH. With stirring add 6 M HCl dropwise until the mixture is acidic.
6. Test the mixture with litmus or pH paper to determine when it is acidic. Note: Follow the direction of your TA at the demo to use the correct way of using this technique.
7. When it is acidic, stir the mixture for 10-20 seconds and note the result (soluble or insoluble) in the table.
8. Repeat this experiment using ethyl 4-aminobenzoate and the same three solvents and record the results.
9. Now take the tube containing ethyl 4-aminobenzoate and 1.0 M HCl. with stirring, add 6 M NaOH dropwise until the mixture is basic.
10. Test the mixture with litmus or pH paper to determine when it is basic. Stir the mixture for 10-20 seconds and note the result.

Note: Grind up the the solid solutes for part A into powder.

References
Introduction to Organic Laboratory Techniques: A Microscale Approach ^{3rd} ed.;Pavia Lampman, Kriz, and Engel.

Safety
1. If exposure to ANY of the reagents occurs, be sure to wash hands and any other affected areas thoroughly with plenty of soap and water.
2. 6M HCl and 6 M NaOH are corrosive. So, prevent contact with eyes, skin, or clothing.
3. Ether is highly flammable: No flames should be allowed in the laboratory when this experiment is being performed.

Disposal
Dispose of the test tubes and pipets in the **" broken glass" bin**, after performing a final rinse with small amount of acetone (0.5 mL).
Dispose of the contents of test tubes in the "Organic Liquid Wastes" receptacle.
Dispose of any excess liquid reagents in the "Organic Liquid Wastes" receptacle.
Dispose of any excess solid reagent in the "Organic Liquid Waste" receptacle after dissolving in a small amount of acetone.

Postlab Questions
1. For each of the following pairs of solutes and solvent, predict whether the solute would be soluble or insoluble. After making your predictions, you can check your answers by looking up the compounds in The Merck Index or the CRC Handbook of Chemistry and Physics or Chem Finder in the ethernet. If the substance has a solubility greater than 40 mg / mL, you may conclude that it is soluble.

    a. Malic acid in water
    b. Naphthalene in water.
    c. Amphetamine in ethyl alcohol
    d. Aspirin in water

Note: For full credit you need to draw structural formulas for the above solutes.

2. Predict whether the following pairs of liquids would be miscible or Immiscible:
    a. Water and methyl alcohol
    b. Hexane and benzene
    c. Methylene chloride and benzene
    d. Water and toluene

Infrared Spectroscopy—Exp#14a

Background

Infrared spectroscopy is used by organic chemists to identify the types and molecular environments of functional groups in organic molecules.

The following figure indicates the position of IR radiation in electromagnetic spectrum. The useful IR (used by the organic chemist) covers the wavelength between approximately: 2.5×10^{-6} to 2.5×10^{-5} m in electromagnetic spectrum.

The Electromagnetic Spectrum

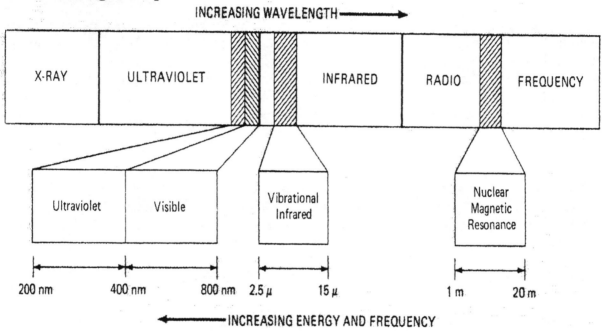

Scale Conversion Examples

$v = 1/\lambda = 1/2.5 \times 10^{-4}$ cm $= 4000$ cm^{-1}

$v = 1/\lambda = 1/2.5 \times 10^{-3}$ cm $= 400$ cm^{-1}

Theory

IR spectra are measured in units of frequency or wavelength. The wavelength is measured in micrometers or microns, μm (1 μm = 1 x 10^{-6} m). The positions of absorption bands are measured in frequency units called wave numbers v^-, which are expressed in reciprocal centimeters, corresponding to the number of cycles of the wave in each centimeter. IR radiation consists of wavelengths that are longer than those of visible light. It is detected not with the eyes but by a feeling of warmth on the skin. When IR radiation interacts with a molecule, the molecular vibration with a frequency matching that of the radiation increases its amplitude. In other words, it is said that molecule absorbs the IR radiation. You can approximately calculate the frequency of a vibration using Hooke's Law.

Hooke's Law

$$\bar{\nu} = \frac{1}{2\pi c}\sqrt{\frac{k}{\mu}}$$

k ⟶ force constant (stiffness)
μ ⟶ reduced mass

C≡C	C=C	C-C
2150	1650	1200

C-H	C-C	C-O	C-Cl	C-Br	C-I
3000	1200	1100	800	550-500	

$\bar{\nu}$ = frequency in cm^{-1}
c = velocity of light = 3×10^{10} cm/sec
k = force constant in dyne/cm
$\mu = \dfrac{m_1 m_2}{m_1 + m_2}$: masses of atoms in grams

Recall: A signal is only observed in the IR spectrum if the net dipole moment of the molecule changes during the interaction with the electromagnetic radiation. This is very likely for groups, which already possess a significant dipole moment (e.g., C-O, C-Cl, O-H, etc.), which usually show strong peaks in the IR spectrum. Where as C≡C stretching (symmetric) bands for the following molecule is not observed in IR spectroscopy because the molecule undergoes no net change of dipole moment when it stretches. For example,

$$H_3C\text{———}C\equiv C\text{———}CH_3$$

How to Interpret the IR Spectrum of a Known Compound:
Step 1: Circle all the functional groups associated with the known molecule.
Step 2: Using correlation charts in your textbook and in your handout, assign peaks for each vibrational mode associated with each functional group.

**NOTE: It is *not* possible to interpret *all* the bands in IR spectrum.

IR Spectrum of Cyclopentylmethanol

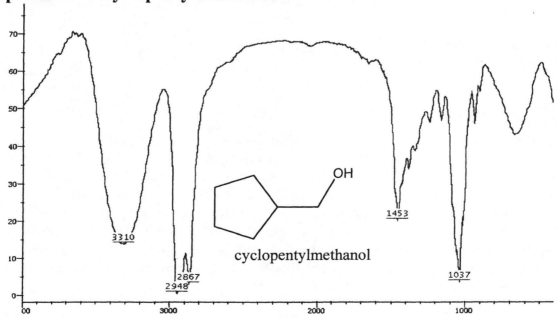

cyclopentylmethanol

IR Spectrum of *trans*-Cinnamaldehyde

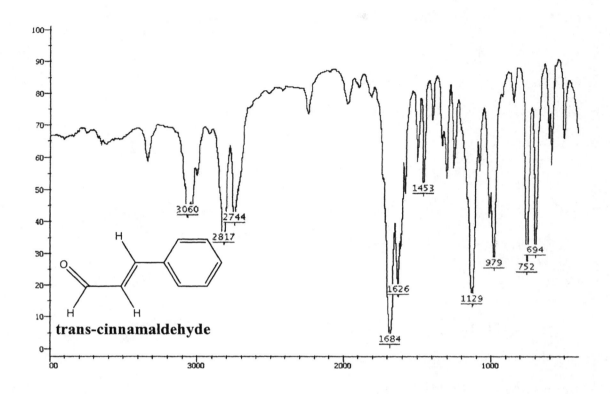

Interpretation of IR Spectrum of Cyclopentylmethanol based on General Correlation Chart

Assignments	Absorption(cm^{-1})	Intensity
Alcohol		
1) -OH stretch	3400–3200	m(broad)
2) C-O stretch	1300–1000	s
ALKANE		
3) C(sp^3)- H stretch	3000–2800	m
4) -CH$_2$- bend	1465	s

Interpretation of IR Spectrum of *trans*-Cinnamaldehyde based on General Correlation Chart

Assignment	Absorption (cm^{-1})	Intensity
Aldehyde		
1) C=O stretch	1740–1720	s
2) (O=C)-H stretch	2900–2800 and 2800–2700	w
Aromatic (aryl)		
3) C(sp^2) - C(sp^2) stretch	1600–1400	m-w
4) C(sp^2)-H stretch	3150–3050	m-w
5) C-H out of plane bend for monosubstituted benzene ring	650–780 (2 peak)	s
Alkene		
6) C(sp^2)-C(sp^2) stretch	1680–1600	s
7) C(sp^2)-H stretch	3100–3000	m-w
8) C-H out of plane bend for disubstituted trans alkene	890–1020	m
		s

- ✓ Now, knowing the above information from the correlation charts, assign peaks on the IR spectra of the compounds shown on pages 56 and 57.

Sample Preparation of Solids—KBr Pellet

1. The sample to be analyzed *must* be as dry as possible. If the sample is not completely dry, a small amount can be placed on a watch glass and placed in a drying oven for 15 minutes.
2. The KBr salt must be kept in a desiccator to prevent water absorption.
3. A clean, dry spatula should be used to obtain a small amount of the KBr (~tip of spatula) to place in a 10 mL beaker.
4. A very small amount of the dry solid sample is added to the beaker containing the KBr sample. ****NOTE**: *The ratio of solid sample to KBr pellet should be about 1:10 to ensure a good KBr pellet formation*.
5. The flat part of the spatula is then used to grind the two solids together until a fine powder is obtained.
6. Obtain a die-press kit box from the instructor.
7. Using the die-press depicted in Figure 16, use the following instructions to create your finalized KBr pellet.

8. Insert the lower piston into the vise and gently screw it in. The nut is then placed on the lower piston so that the piston passes halfway through the nut. The KBr and sample mixture is scooped into the nut. **NOTE:** *Use only enough of the mixture to just cover the piston the nut sits upon.*
9. Gently tap the press on the bench top to distribute the sample evenly on the piston.
10. While keeping the die-press held upright, gently screw the two pistons into the vise until the sample is firmly sandwiched between them.
11. Use the gooseneck wrenches to apply more pressure to the sample while carefully tightening the screws. **CAUTION:** *Ensure that the wrenches are straightened within the press before attempting to tighten the screws. Failure to do so will damage the wrenches!!*

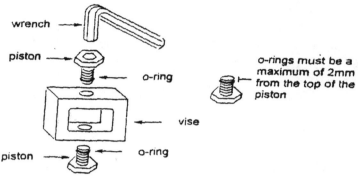

Figure 16: Die-Press

12. Apply pressure to the wrenches for 3 minutes, loosen the die-press using the gooseneck wrenches and practicing the precautions and instructions from step 11.
13. While stabilizing the nut containing your sample, gently unscrew the pistons using your fingers once the pistons have been loosened.
14. Remove the nut containing your KBr/sample pellet. The pellet should be translucent yet stable within the nut (not falling apart or crumbling out of the nut).
15. If a good pellet has been obtained, return the die-press kit to the instructor. If a good pellet was *not* obtained, please refer to the list of common problems below and reconfigure your steps to obtain a proper pellet formation.
16. Continue with the instructions for operating an IR spectrometer as directed on page 59.

**Common problems that may result in an unsatisfactory pellet formation
- The KBr mixture may not have been ground finely enough—particle size is too big.
- The sample of KBr used was not **completely** dry.
- An improper ratio of sample to KBr may have been used (was not a 1:10 ratio)—too much sample for the amount of KBr used.
- The pellet may be too thick—too much of the sample and KBr mixture was put into the die-press.
- The sample itself may have too low of a melting point. Samples having a melting point that is too low are difficult to dry as well as tend to melt under the pressure of the die-press. Under this condition, the sample may need to be dissolved in a Nujol solvent or a solution of carbon tetrachloride. The instruction for preparing a solid sample—Nujol Mull method should be followed.

Sample Preparation of Solids—Nujol Mull (NaCl Salt Plates)

This method is used under the following conditions:
- An adequate KBr pellet **cannot** be obtained.
- The sample is insoluble in CCl_4 or $CHCl_3$.
- The sample has a low melting point (e.g., <50 °C).

1. Grind approximately 5 mg of the solid sample in a 10 mL beaker using a spatula or a mortar with a pestle.
2. Add 2–3 drops of Nujol mineral oil to the sample and continue to grind the mixture to a very fine dispersion.
3. Obtain two NaCl salt plates from the dessicator.
4. *Under the fume hood:* Ensure the cleanliness of the salt plates by adding a few drops of CH_2Cl_2 to one side of the plates, then gently rub in a circular motion with a paper towel to clean. Flip over the plates and clean that side as well. ****CAUTION:** *NEVER USE WATER OR ACETONE TO CLEAN THE SALT PLATES!! THIS WILL DAMAGE THEM!!*
5. Make sure all of the CH_2Cl_2 has evaporated before applying your sample to the salt plates.
6. Use a rubber policeman or a polyethylene pipette to place 1 drop of the suspension between two salt plates.
7. Continue with the instructions for operating an IR spectrometer as directed on page 62.
8. Once an IR spectrum is obtained, follow the instructions outlined in step 4 to clean the plates. Return the clean plates to the desiccators.

Sample Preparation of Liquids—NaCl Salt Plates

1. Obtain two NaCl salt plates from the desiccator.
2. *Under the fume hood:* Ensure the cleanliness of the salt plates by adding a few drops of CH_2Cl_2 to one side of the plates, then gently rub in a circular motion with a paper towel to clean. Flip over the plates and clean that side as well. ****CAUTION:** *NEVER USE WATER OR ACETONE TO CLEAN THE SALT PLATES!! THIS WILL DAMAGE THEM!!*
3. Place 1–2 drops of a liquid sample between the NaCl salt plates.
4. The combining of the plates should cause the liquid to spread out as a "film" sandwiched between the two salt plates. ****NOTE:** *If more than one or two drops is used of the sample, the result may be an inaccurate IR spectrum.*
5. Continue with the instructions for operating an IR spectrometer as directed on page 62.
6. Once an IR spectrum is obtained, follow the instructions outlined in step 2 to clean the plates. Return the clean plates to the desiccators. ****NOTE:** *Be sure to handle the salt plates by their edges.*

Experimental Procedure—Infrared Spectroscopy
1. Prepare the following IR samples for analysis using the procedures outlined above:
   a. Liquid sample of isopentyl alcohol using NaCl salt plates.
   b. Solid KBr pellet of benzoic acid.
2. Run the IR spectrum of each sample using the instrument instructions provided on page 62.
3. Identify all possible absorption peaks in the two IR spectra you have obtained.
➢ **For check out, you need to show the following to your instructor.**
   - The IR spectra of the two above samples (fully analyzed).
   - Show your work (problems 1–5) and obtain your instructor's signature.
   - Your spectroscopy handout is due next week.
➢ **Remember, no experimental write-up is required for this lab. Just work on your spectroscopy handout.**

References
Organic Chemistry 7th ed.; John McMurry.
Introduction to Organic Laboratory Techniques: A Microscale Approach 3rd ed.; Pavia, Lampman, Kriz, and Engel.

Safety
1. **Clean (using soap for glassware) and rinse (using distilled water for final rinse) ALL glassware and utensils—EXCEPT the die-press kit apparatus and the salt plates—PRIOR to returning them to your lab drawer.**
2. **Clean the die-press apparatus and nut with a small amount of acetone and a paper towel. **NOTE**: *Remove the O ring BEFORE applying the acetone.***
3. **Clean the salt plates by placing them face up on a paper towel, applying a small amount of CH_2Cl_2 to the plates, and then gently rubbing the plates with the paper towel in a circular motion. Repeat this procedure for the other side of the plates.**
4. **If exposure to ANY of the reagents occurs, be sure to wash hands and any other affected areas thoroughly with plenty of soap and water.**

Disposal
Dispose of the solid pellet in the "Organic Solid Wastes" receptacle. **NOTE:** *Use a wooden applicator stick to remove the pellet from the nut* **DO NOT USE A METAL SPATULA!!**
Dispose of the liquid reagents in the "Organic Liquid Wastes" receptacle.
Dispose of any disposable pipettes and/or test tubes in the "broken glass" bin.

Postlab Questions
There are **NO** postlab questions for this experiment; however, the spectroscopy handout must be completed entirely in order to receive full credit. **SHOW ALL WORK!!**

A Quick Guide to Operating the IR Solution Spectrum

Obtaining the initial spectrum:
1. Initialize the IR solution by double-clicking on the **[IR solution]** icon on the desktop.
2. To measure the background, select the parameters you want to scan and to be run from the right panel of the screen (e.g., range of your wave number: 650–4000 cm^{-1}, and scan #: 16, and resolution: 2).
3. Make sure the mode of measurement is in % transmittance and *not* absorbance.
4. A dialog box will appear stating: **[found background data from previous sessions, remove the marked data?]**. Click on **yes**.
5. Make sure that the sample compartment is clear of any other samples or debris.
6. On the measure bar menu, click on the **[BKG]** tab. A dialog box will appear stating: [the sample compartment to be prepared for a background scan]. Click on **OK**.
7. The background scan will take place. **DO NOT CLICK ON ANYTHING WHILE THE SCAN IS COMMENCING!!** The numbers in the left-hand corner will cease to move when the scan is finished. Once the scan is complete, you will be able to view the background spectra obtained.
8. To measure your sample, open the sample compartment and place your sample in the holding chamber. ****NOTE:** *Make sure the laser beam passes through the sample to be measured!!*
9. Click on the **[measure]** tab. The FTIR will begin to take your sample measurement. Once the 16 scans have run, the data obtained will be ready for further analysis. *Once again refer to the bottom left corner of your screen to monitor the scan completion.*
10. At this time, you will be able to view **2** spectra on the screen.
11. ****NOTE:** *YOU SHOULD **NOT** SAVE YOUR FILE!!*

Manipulation of spectra data:
1. Click on the **[manipulation]** tab on the main menu at the top of the screen. Scroll down to **[peak table]**.
2. On the right side of the screen, set the noise to 0.1 and press the **[enter]** key. This will display two spectra above and below the window panel.
3. To obtain a darker trace spectrum: (1) Click on **graph preferences.** (2) Click on **coloring.** (3) Click on **active object color.** (4) Click on **black.** (5) Click on **OK.** (6) Click on '**OK**' again. Proceed with the next step for printing.
4. To delete any unwanted peaks, select the unwanted peak wave # from the scroll down menu displayed in the "manual peak pick panel" and then click the **[delete]** tab.
5. Click on the **[OK]** tab at the bottom of the right panel.
6. To print your finalized spectra, click on **[file]** on the top of the menu screen. Scroll down to "preview" and select.
7. A dialog box will appear asking you to select a template. Select the template labeled "organic chemistry lab," then click **[open]**.
8. Your sample spectrum with a table of data (e.g., peak wave #, peak intensity, peak correlation, intensity, etc.) will be displayed beneath the graph.
9. Observe the IR spectrum of your sample and click on the [print] tab in the upper left corner of the screen.
10. To exit, scroll down on the file menu bar, then select "exit." The IR solution icon will once again be displayed.

IR Absorption Frequencies Correlation Chart

Vibration Type		Frequency (cm^{-1})	Intensity
C-H stretches			
C(sp^3)-H stretch	Alkane	3000–2850	m-s
C(sp^2)-H stretch	Alkene	3100–3000	m
C(sp^2)-H stretch	Aromatic	3150–3050	s
C(sp)-H stretch	Alkyne	3300	s
(O=C)-H stretch	Aldehyde	2900–2800 and 2800–2700	w
CH$_n$ bends			
-CH$_3$ bend	Alkane	1450 and 1375	m
-CH$_2$- bend	Alkane	1465	m
=C-H bend	Alkene	1000–650	s
-C-H o.o.p bend	Aromatic	1000–700	s
C-C stretches			
C(sp^3)-C(sp^3) stretch	Alkane	variable	-
C(sp^2)-C(sp^2) stretch	Alkene	1680–1600	m-w
	Aromatic	1600–1400	m-w
C(sp)-C(sp) stretch	Alkyne	2250–2100	m
		no IR peak when symmetrical	
C=O stretches			
Amide		1700–1640	s
Carboxylic Acid		1725–1700	s
Ketone (acyclic)		1725–1705	s
Aldehyde		1740–1720	s
Ester		1750–1730	s
Acyl Halide		1800–1760	s
Anhydride		1810 and 1760	s
C-O stretches			
Alcohol		1150–1050	s
Ether, Ester & Anhydride		1300–1000	s
Carboxylic Acid		1320–1210	s
C-N stretches			
1° Aliphatic Amine / Amide		1140–1070	m
2° Aliphatic Amine / Amide		1190–1130	m-s
1° Aromatic Amine / Amide		1330–1260	s
2° Aromatic Amine / Amide		1340–1250	s
Nitrile		2260–2210	m
N-H stretches			
1° Amine		3500–3300 (doublet)	m
2° Amine		3500–3300 (singlet)	m-w
1° Amide		3350 and 3180	m
2° Amide		3300 (singlet)	m
N-H bends			
1° and 2° Amines		1640–1560	m
1° and 2° Amides		1640–1550	m
N=O stretches			
Nitro group		1600–1500 and 1400–1300	s
O-H stretch			
O-H (H bonded)	Alcohol and Phenols	3400–3200	s (broad)
O-H (free)	Alcohol and Phenols	3650–3600	s
O-H	Carboxylic Acid	3300–2500	s (broad)
Alkyl Halide stretches			
C-F		1400–1000	s
C-Cl		800–600	s
C-Br		600–500	s
C-I		500	s
Ketone stretches (specific)			
Cyclic:	3-membered ring	1850	s
	4-membered ring	1780	s
	5-membered ring	1745	s
	6-membered ring	1715	s
	7-membered ring	1705	s
α, β – Unsaturated		1685–1665	s
Aryl		1700–1680	s

NOTE: s=strong, m=medium, w=weak.

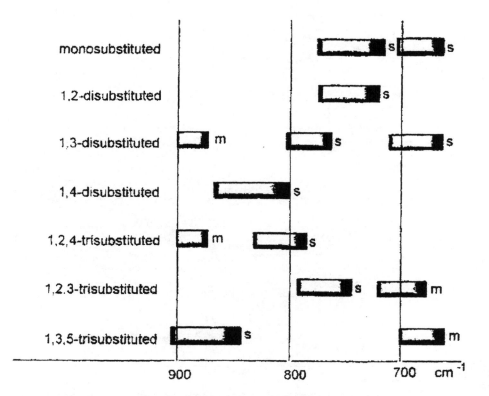

C-H OUT-OF-PLANE BENDS (O.O.P.B) AND OVERTONE/COMBINATION BANDS FOR DIFFERENT SUBSTITUTED AROMATIC RINGS.

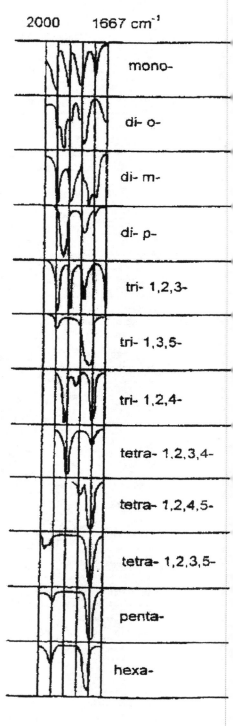

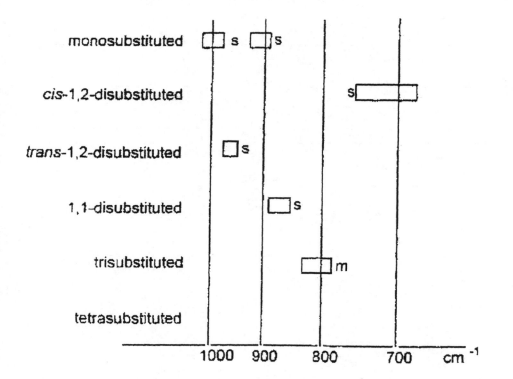

THE C-H OUT-OF-PLANE BENDING VIBRATIONS FOR SUBSTITUTED ALKENES.

NMR Spectroscopy—Exp#14b

Background
Proton (^{1}H) nuclear magnetic resonance (NMR) spectroscopy is used by organic chemists worldwide to identify the chemical structures of organic compounds. By utilizing this method, the number, kind, and relative locations of the hydrogen atoms of a compound can be located. (^{13}C) NMR, in a similar manner of the (^{1}H) NMR, is a method by which the relative location, number, and different kinds of carbon atoms in a molecule can be obtained. The information obtained from these two forms of NMR spectroscopy can then be analyzed to identify a wide range of functional groups, the general connectivity of the atoms, as well as the structural formula of the compound in question. Furthermore, by combining nuclear magnetic resonance with IR spectroscopy, the structural connectivity of most compounds can be simplified to an exact chemical structural and characterization of the organic compound.

Goal
In this laboratory you will learn how to interpret ^{1}H NMR spectra of known compounds and unknown compounds. The following discussion provides a general overview of NMR spectroscopy. *You are required to consult your organic chemistry textbook and lecture notes for more detailed discussions of the concepts presented here. Please bring your textbook to lab.*

Theory
Similar to electrons, atomic nuclei have the ability to "spin" either clockwise or counterclockwise on an axis. Certain nuclei (i.e., ^{1}H and ^{13}C) behave as miniature magnets in the presence of an applied magnetic field. Within a magnetic field, these nuclei will either align themselves with the field or against it. Under normal conditions, most nuclei will align with an applied magnetic field instead of opposing it. However, if a proper amount of energy is introduced in the form of radiofrequency waves, the magnetic nuclei will absorb this energy and flip whereby a nuclear spin transition will occurs. This causes the nuclei to spin in opposition to the field. Traditional NMR spectrometers supply energy at a constant radiofrequency while varying the strength of the applied magnetic field. The UNLV chemistry department utilizes two NMR spectrometers with each operating at its own constant radiofrequency—one at 60 MHz and one at 400 MHz. In ^{1}H NMR, hydrogens in different electronic environments will **resonate**, or spin flip, at different magnetic field strengths while the radiofrequency is held constant. These different resonances are recorded as peaks in an NMR spectrum. The energy for each peak is unique for each different type of hydrogen present in the molecule. Sometimes the difference in energy between these resonance peaks is small and they overlap with one another. The relative area under each peak, obtained by integration, corresponds to the relative number of hydrogens (hereafter referred to as protons) giving rise to that peak.

Chemical Shift—Where Do the Peaks Appear in an NMR Spectrum?

The position of a resonance peak (or the energy of it) in a ^{1}H NMR spectrum is known as the *chemical shift*. Chemical shift (δ) is defined as:

δ (ppm) = observed shift from TMS (Hz)/radiofrequency (MHz)

Chemical shift values are reported in parts per million (ppm) relative to a *reference* peak. Tetramethylsilane (TMS) is used as the reference material because it gives rise to a very large peak (corresponding to 12 protons) at a position that is **upfield** from proton resonances observed for typical organic compounds. The TMS reference peak is set at 0.00 ppm, and most other resonance peaks appear **downfield** from the TMS peak, usually in the range of 0.00–12.00 ppm. Table 1 indicates general chemical shift values for different types of protons. The chemical shift value for a given proton is strongly influenced by the electronegativity of the atom it is directly connected to as well as the other atoms that are in close proximity. *In general, chemical shift increases as the electronegativity of the attached atom increases.* For example, in Table 1, the protons of a methyl group attached to another carbon (C-CH$_3$) resonate at a *smaller* chemical shift value (0–2 ppm) compared to a methyl group attached to a more electronegative atom, such as oxygen (O-CH$_3$, 3–5 ppm). Please consult your organic chemistry textbook and lecture notes for a more detailed discussion of chemical shift trends.

Table 1. A General Correlation Chart for Proton Chemical Shift Values

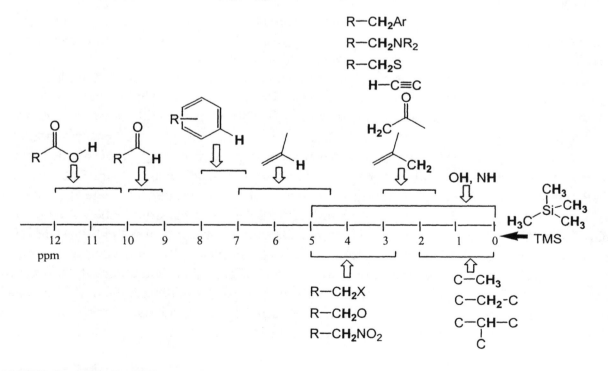

**NOTE: R = H or alkyl group.

Proton Chemical Shifts Correlation Chart

Type of Proton	Structure	Chemical Shift (ppm)
Primary aliphatic	R-CH_3	0.8–1.0
Secondary aliphatic	R_2-CH_2	1.2–1.5
Tertiary aliphatic	R_3-CH	1.4–1.7
Vinyic	C=C-**H**	4.6–5.9
Allylic	C=C-CH_3	1.6–1.9
Acetylenic—aliphatic	R-C≡C-**H**	2.3–2.5
Acetylenic—aromatic	Ar-C≡C-**H**	2.8–3.1
Aromatic	Ar-**H**	6.0–8.5
Benzylic	Ar-CH_3	2.2–3.2
Alcohol	R-CH_2-OH	3.4–4.0
Hydroxylic	R-CH_2-O**H**	1.0–5.5
Phenolic	Ar-OH	4.0–12.0
Enolic	C=C-OH	15.0–17.0
Amine—aliphatic	R-NH_2	0.6–2.5
Amine—aromatic	Ar-NH_2	3.0–4.5
Ether—aliphatic	R-O-CH_3	3.2–3.5
Ether—aromatic	Ar-O-CH_3	3.7–4.0
Ketone—aliphatic	R-(C=O)-CH_3	2.1–2.4
Ketone—aromatic	Ar-(C=O)-CH_3	2.4–2.6
Aldehyde—aliphatic	R-(C=O)-**H**	9.0–10.0
Aldehyde—aliphatic	R-CH_2-(C=O)-H	2.1–2.4
Aldehyde—aromatic	Ar-(C=O)-**H**	9.7–10.3
Ester—aliphatic	R-(C=O)-O-CH_3	3.7–4.1
Ester—aliphatic	H_3C-(C=O)-O-R	2.0–2.2
Ester—aromatic	Ar-O-(C=O)-CH_3	2.0–2.5
Ester—aromatic	Ar-(C=O)-O-CH_3	4.0–4.2
Carboxylic acids—aliphatic	R-(C=O)-OH	10.4–12.0
Carboxylic acids—aliphatic	R-CH_2-(C=O)-OH	2.0–2.6
Carboxylic acids—aromatic	Ar-(C=O)-OH	10.4–12.0
Amide—aliphatic	R-(C=O)-NH_2	5.5–7.5
Amide—aliphatic	R_2N-(C=O)-CH_3	1.8–2.2
Nitrile	R-CH_2-C≡N	2.0–2.3
Nitro	R-CH_2-NO_2	4.2–4.6
Fluoride	R-CH_2-F	4.0–4.5
Chloride	R-CH_2-Cl	3.0–4.0
Bromide	R-CH_2-Br	2.5–4.0
Iodide	R-CH_2-I	2.0–4.0
Thiols—aliphatic	R-SH	1.0–2.0
Thiols—aromatic	Ar-SH	3.0–4.0

Carbon-13 Chemical Shifts Correlation Chart

Type of Carbon	Structure	Chemical Shift (ppm)
Primary aliphatic	R-CH₃	10–15
Secondary aliphatic	R₂-CH₂	15–25
Tertiary aliphatic	R₃-CH	25–50
Vinylic	R-CH=CH₂	115–120
Vinylic	R-CH=CH₂	125–140
Acetylenic	RC≡CH	65–70
Acetylenic	RC≡CH	75–85
Aromatic	Ar-H	125–150
Benzylic	Ar-CH₃	125–140
Alcohol	R-CH₂-OH	50–65
Amine	R-CH₂-NH₂	35–50
Ether	R-CH₂-O-R'	50–90
Ketone	R-(C=O)-CH₃	30
Ketone	R-(C=O)-CH₃	205–215
Aldehyde	R-(C=O)-H	190–200
Ester	R-O-(C=O)-CH₃	20
Ester	R-O-(C=O)-CH₃	170–175
Carboxylic acid	R-(C=O)-OH	175–185
Amide	R-(C=O)-NH₂	150–180
Amide	R-(C=O)-NR'₂	150–170
Nitrile	R-C≡N	115–130
Chloride	R-CH₂-Cl	40–45
Bromide	R-CH₂-Br	25–35

Chemical Equivalence—How many different proton resonances will a compound have?
How many different proton resonances will be observed for a given compound? In order to determine this you must first understand the concept of chemical equivalence.
- Protons in the same electronic environment will resonate at the same chemical shift value.
- This means that in symmetrical molecules, all protons are *chemically equivalent* and will exhibit only one resonance peak in the ¹H NMR spectrum.
- Four examples are provided below.

SYMMETRICAL MOLECULES—ALL PROTONS ARE CHEMICALLY EQUVALENT

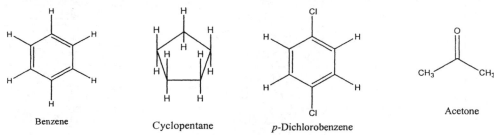

Benzene   Cyclopentane   *p*-Dichlorobenzene   Acetone

In the next set of examples, the molecules are no longer symmetrical.
- Each molecule has **two sets of equivalent protons** and therefore will exhibit two resonance peaks in their respective ¹H NMR spectra.

NONSYMMETRICAL MOLECULES—ALL PROTONS ARE **NOT** CHEMICALLY EQUIVALENT

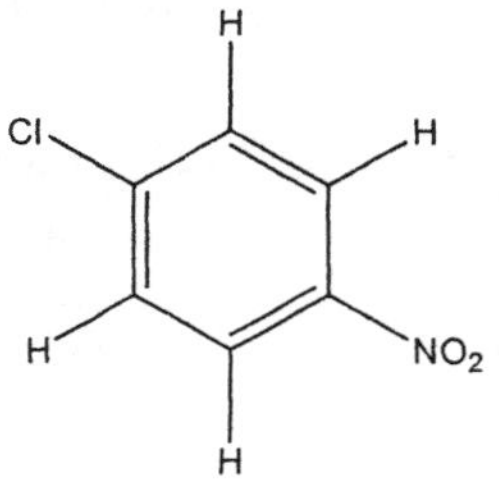

p-Chloronitrobenzene 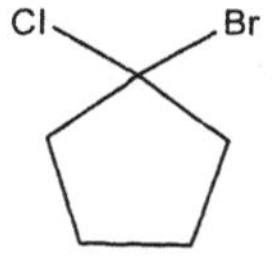   1-Bromo-1-chlorocyclopentane 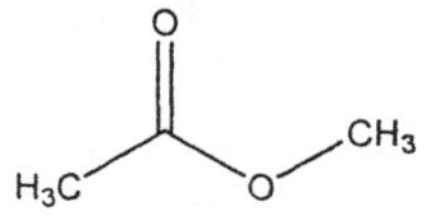   Methylacetate

Practice Problem 1: For each of the following molecules, how many different proton resonance peaks do you expect to observe? Once you have determined *how many* resonance peaks you expect to see in each ^{1}H NMR spectrum, determine the approximate *chemical shift* value (in ppm) for each resonance peak.

Calculate the number of different types of protons in the molecules shown below:

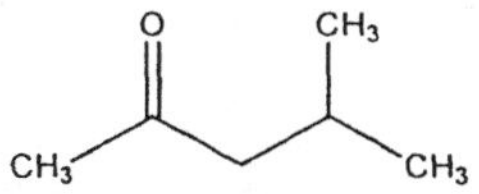

I 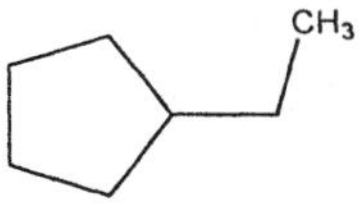   II 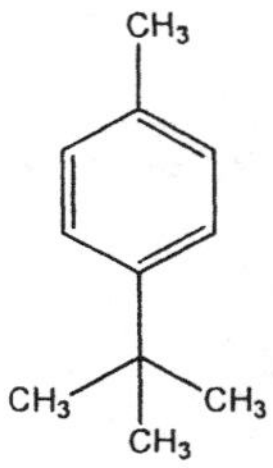   III

You should have calculated:
- I = 4 types of protons
- II = 5 types of protons
- III = 4 types of protons

The (N + 1) Rule—What will the resonance peak look like? Singlet, doublet, triplet, etc.?

Once you have determined *how many* proton resonance peaks you should see for a particular compound (evaluation of chemical equivalency) and *where* they should appear in the spectrum (chemical shift), you can also use ^{1}H NMR to determine *how the groups of atoms are connected* to each other. Each proton resonance peak will be "split" into a smaller group of peaks because it is influenced by other protons near it in the molecule. This phenomenon, called *spin-spin splitting,* can be explained empirically by the *(N + 1) Rule:*

One type of chemically equivalent proton (H_a) will "sense" the number of (N) protons of a different type (H_b) on the neighboring carbon atom(s). The result is a resonance peak for each type of proton, which is split into N + 1 components.

Several examples of the **(N + 1) Rule** are provided on next page.

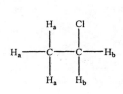

Proton H_b "senses" the presence of protons H_a ($N = 2$), so the resonance peak for H_b will be split into a group of $(2+1) = 3$ peaks or a *triplet*. Protons H_a "sense" the presence of proton H_b ($N = 1$), so the resonance peak for protons H_a will be split into a group of $(1+1) = 2$ peaks or a *doublet*.
At approximately what chemical shift values will these two peaks appear?

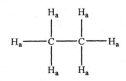

Proton H_b "senses" the presence of protons H_a ($N = 3$), so the resonance peak for H_b will be split into a group of $(3+1) = 4$ peaks or a *quartet*. Protons H_a "sense" the presence of proton H_b ($N = 1$), so the resonance peak for protons H_a will be split into a group of $(1+1) = 2$ peaks or a *doublet*.
At approximately what chemical shift values will these two peaks appear?

Protons H_b "sense" the presence of protons H_a ($N = 3$), so the resonance peak for protons H_b will be split into a group of $(3+1) = 4$ peaks or a *quartet*. Protons H_a "sense" the presence of protons H_b ($N = 2$), so the resonance peak for protons H_a will be split into a group of $(2+1) = 3$ peaks or a *triplet*.
At approximately what chemical shift values will these two peaks appear?

All the protons in ethane are chemically equivalent and when no splitting occurs, a single resonance peak called a *singlet* will be observed.
At approximately what chemical shift values will this peak appear?

Practice Problem 2: For each of the following molecules use the $(N + 1)$ Rule to predict the splitting pattern of all resonance peaks and draw the predicted ^{1}H NMR spectrum.

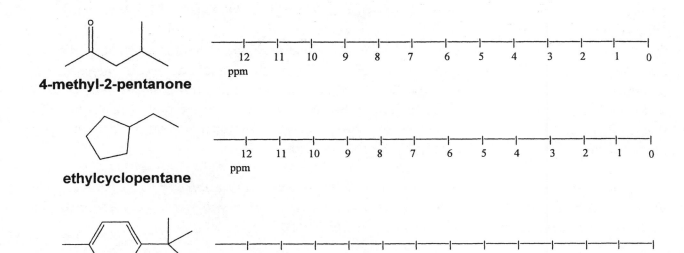

74

Integration of NMR Peaks—How do you determine how many protons are associated with each resonance peak?

The NMR spectrometer performs an integration of each peak or group of peaks in the spectrum. These integrals correspond to relative areas and will appear above each peak or set of peaks. Many times the raw integral will appear with numerical integration values, but the relative areas can also be calculated manually. Integration becomes very important when you are not sure of the peak assignment. Let's look at an example. The ^{1}H NMR spectrum of cyclopentylmethanol is shown in Figure 17. This molecule has five types of protons in different chemical environments (labeled a–e in the figure), so we expect to observe five different resonance peaks. Based on the (N + 1) Rule, the following splittings should be observed: H_a (triplet), H_b (multiplet), H_c (multiplet), H_d (doublet), and H_e (broad singlet). Looking at the spectrum, note that there is a doublet located at about 3.5 ppm that must correspond to protons H_d. The remaining peaks in the spectrum are overlapping and it is difficult to assign them without more information. Integration becomes very important at this point. If the doublet in the spectrum is assigned to protons H_d and assigned a numerical value corresponding to 2 (the number of H_d protons), then this peak can be used as the reference peak for calculating the number of protons corresponding to the remaining integrals in the spectrum. There are two ways to do this:

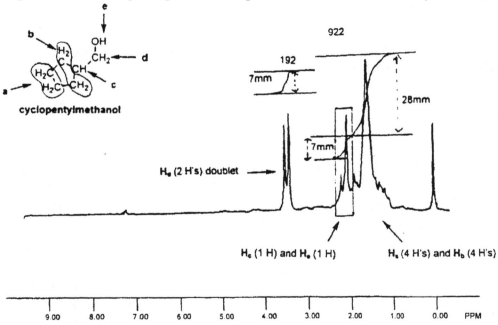

Figure 17: ^{1}H NMR spectrum of cyclopentylmethanol

1. **Use the numerical integration values provided.** In the cyclopentylmethanol example, an integral value of 192 appears for the H_d doublet, which corresponds to 2 protons. To use this value as the reference (normalize), divide 192 by itself and multiply by 2: (192/192) x 2 = 2. Now normalize the other integral values by dividing each by the reference value and multiplying by 2. In this example there is only one additional integral: (922/192) x 2 = 9.6, or approximately 10. The total number of protons in

cyclopentylmethanol is 12, which corresponds to the calculations obtained from the numerical integration values.

2. **Calculation of integral values manually.** The numerical integration values generated by the instrument software are not always accurate, and therefore it is important to know how to calculate values from the raw integrals. Referring back to the example spectrum for cyclopentylmethanol in Figure 16, parallel lines are drawn through the horizontal tails of each integral, as shown, then a ruler is used to measure the distance in mm between the two lines. This number is the integral value that can be used to calculate the number of protons under each integral as described above for computer-generated values. The difference is that the raw integral can be separated into smaller parts to obtain a more detailed assignment of overlapping peaks. In the example, an integral value of 7 mm is measured for the H_d doublet. So, if we normalize, (7/7) x 2 = 2. By normalizing the other integrals, we are able to separate some of the peaks out from the overlapping peaks in the region 1.0–2.0 ppm. The smaller integral measures 7 mm and thus the peaks under that integral correspond to 2 protons. The larger integral measures 28 mm, so normalization gives (28/7) x 2 = 8 protons. This gives a total of 12 protons in the compound. We also have gained additional information for assigning the overlapping peaks.

Practice Problem 3: Assign all of the proton peaks corresponding to the ^{1}H NMR spectrum shown below for phenacetin.

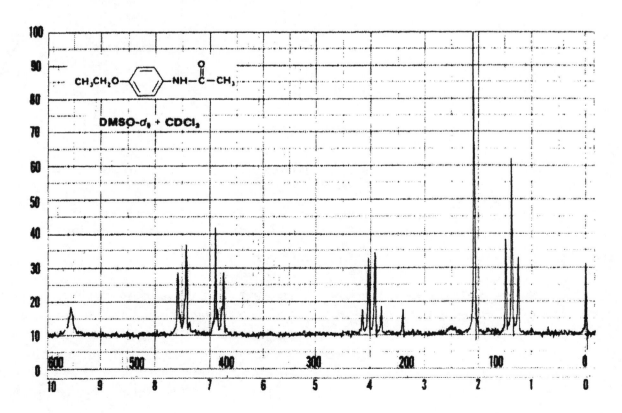

Calculate the number of protons for each of the peaks in the ¹H NMR spectrum for benzyl acetate and assign all of the peaks.

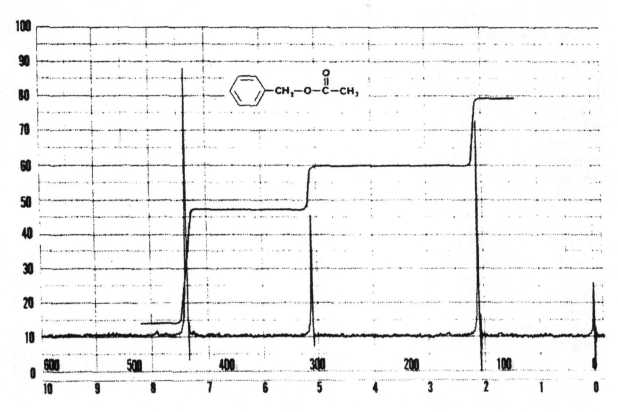

References
Organic Chemistry 7th Edition; John McMurry.
Introduction to Organic Laboratory Techniques: A Microscale Approach 3rd ed.; Pavia, Lampman, Kriz, and Engel.
Laboratory Investigations in Organic chemistry; David C. Eaton.

Safety
There are no real safety precautions for this laboratory experiment.

Disposal
There are no disposal procedures for this laboratory experiment.

Postlab Questions
There are no postlab questions for this lab. Completion of the entire NMR handout/packet is required for full credit for this lab report. **SHOW ALL WORK!!**

Williamson Ether Synthesis—Exp#15

Background
The simplest method for synthesizing ether is via an S_N2 reaction between an alkoxide anion acting as a nucleophile and ethyl p-toluenesulfonate as a substrate.

Theory
The formation of ether can be accomplished by first introducing the water-soluble phenoxide salt to an aqueous solution. Since the alkyl p-toluenesulfonate is not soluble in an aqueous solution, alkyl p-toluenesulfonate cannot be added alone—two immiscible layers will form, preventing the phenoxide and alkyl p-toluenesulfonate molecules from reacting with one another. The addition of the substrate, ethyl p-toluenesulfonate, must be counteracted by tetrabutylammonium bromide *(a phase-transfer catalyst)* for the desired reaction between the reactants to occur.

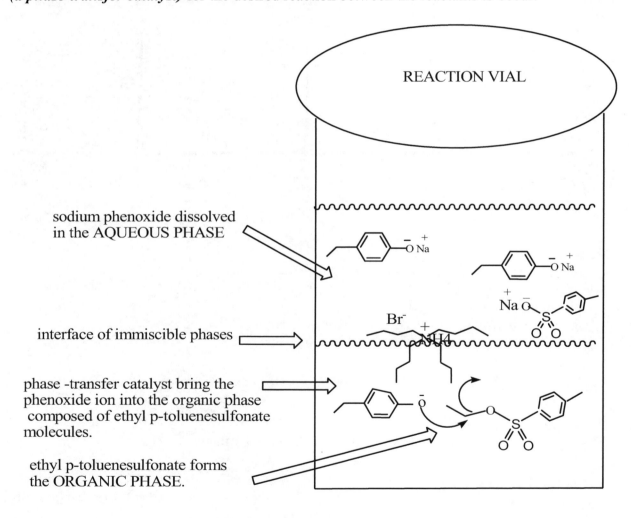

Figure 18: A phase-transfer catalyst brings the two immiscible reactants together.

The tetrabutylammonium bromide forms an "ion pair" with the phenoxide anion. This association allows the long alkyl chains of the tetrabutylammonium cation to stabilize the phenoxide anion within the organic phase, which contains ethyl p-toluenesulfonate. During this reaction, the catalyst does **_NOT_** react with the two reactants; rather it acts only as an aid to bring the two components together for the desired reaction to occur.

Preparation of Alkoxides

An alkoxide nucleophile is **_NOT_** efficiently prepared by the reaction between an alcohol and the base, NaOH. For an S_N2 reaction, the alcohol is a weaker acid than water. This ineffectiveness tends to increase as the size of the alkyl chain is increased. An active metal such as potassium or sodium functions as a more effective method for preparing the desired metal alkoxide compound. Sodium hydride, NaH is even more effective for preparing alkoxides.

In contrast, a phenol is a strong acid when compared to water and therefore the alkoxide of a phenol *(phenoxide)* can be efficiently formed via an acid–base reaction.

General Reaction

Step 1—Formation of the alkoxide ion

$$R-O-H + M \longrightarrow R-O^- M^+ + \tfrac{1}{2} H_2 \uparrow$$

Step 2—S_N2 Mechanism
  **Recall reaction rate: $k [R\text{-}X] [R\text{-}O^- M^+]$

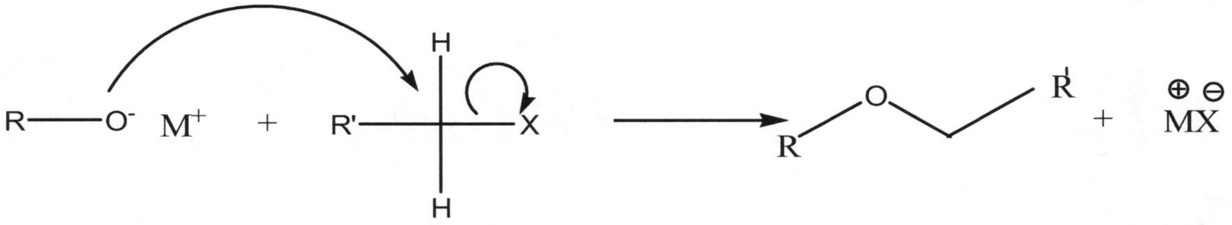

$$R-O^- M^+ + R'-CH_2-X \longrightarrow R-O-CH_2-R' + MX$$

  **$R = 1°, 2°,$ and $3°$ alkyl chain or aromatic ring
    **$R' = 1°$ only
  **$X = Cl^-, Br^-, I^-, TosO^-$

Specific Mechanism (S_N2)

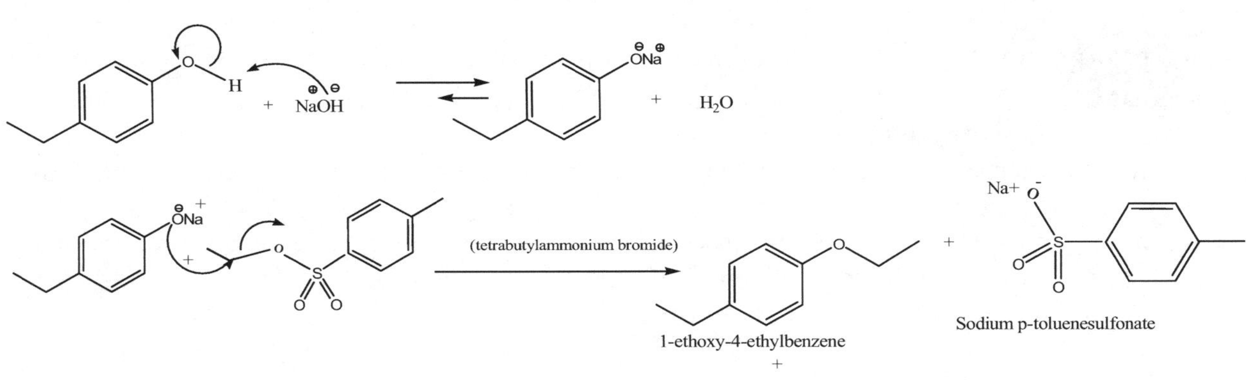

(4-ethylphenol) + NaOH ⇌ (sodium 4-ethylphenoxide) + H_2O

(sodium 4-ethylphenoxide) + ethyl p-toluenesulfonate —(tetrabutylammonium bromide)→ 1-ethoxy-4-ethylbenzene + Sodium p-toluenesulfonate

79

The best substrate or alkyl halide to use for this reaction is primary. The alkoxide and phenoxide anions are not only strong nucleophiles; they are also strong bulky bases. The presence of this strong base, especially with a tertiary alkyl halide, causes the competing E2 elimination reaction to occur instead of the desired S_N2 reaction. For example:

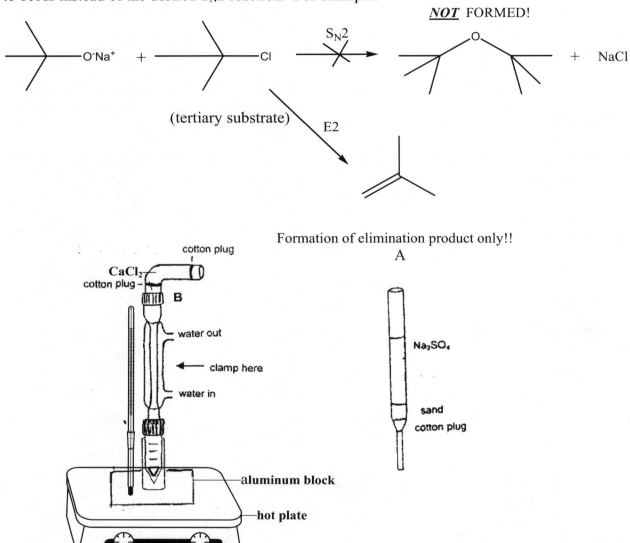

Formation of elimination product only!!

Figure 19 (a) and (b): (a) Micro filtration column (b) Williamson Ether Synthesis setup

Procedure

Formation of the *p*-ethylphenoxide salt:
1. Prior to assembling your Williamson Ether Synthesis setup, be sure to lubricate the end of the condenser that connects to the conical vial using silicon oil. This lubrication process is necessary to prevent fusion of the glassware during the condensation process.
2. Using a calibrated balance, weigh 1.20 mmol of 4-ethylphenol into a clean, dry 5 mL conical vial, then add a spin vane to the vial.

3. Add 1 mL of a 25 % NaOH solution to the vial using the measurement markings on the side of the conical vial.
4. Gently stir the mixture *at room temperature* until all of the solid compound has been dissolved. **<u>NOTE</u>:** *The solution may become viscous and thick during this portion of the experiment—manual mixing with the spatula may be required.*

Formation of the 1-ethoxy-4-ethyl benzene:
1. Once the solids have dissolved, add 0.05 mmol of tetrabutylammonium bromide, the phase-transfer catalyst, to the solution.
2. Then add 1.20 mmol of ethyl p-toluenesufonate to the flask
3. Modify the conical vial with a water-cooled condenser as shown in Figure 19 (b). While continuing to stir the solution, heat the reaction vial to approximately 130 -140 °C using an aluminum block and hot plate. The hot plate setting should be at three. **<u>NOTE</u>:** *The mercury or kerosene reservoir of the thermometer should not be touching the hot plate surface during this heating process. The reservoir should be midway inside the aluminum heating block.*
4. Continue heating the solution in the vial for 1 hour while it is continuously being stirred.

Isolation of the product:
1. Following the one hour reflux process, you should observe two layers, identify each layer and write down your observations in your lab notebook. Then remove the heating source and discontinue the water source to the water-cooling apparatus. Allow the reaction vial to cool to room temperature, *then remove the vial from the condenser.* The reaction mixture may be viscous or it may even be solidified..
2. Next, add 1 mL of water to dissolve the majority of the precipitate formed.
3. An extraction must now be performed in order to isolate the product from the crude reaction mixture.

Figure 20: Demonstration of extraction technique

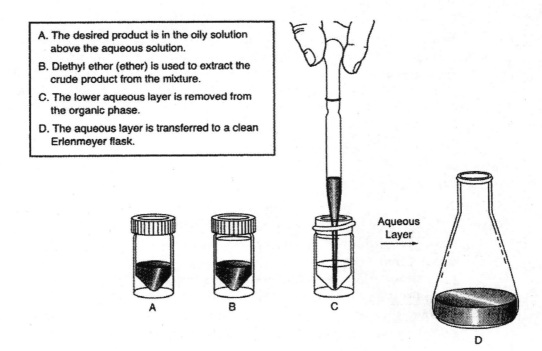

A. The desired product is in the oily solution above the aqueous solution.
B. Diethyl ether (ether) is used to extract the crude product from the mixture.
C. The lower aqueous layer is removed from the organic phase.
D. The aqueous layer is transferred to a clean Erlenmeyer flask.

4. Add approximately 1–2 mL of diethyl ether to contents of the cooled conical vial. *(Why?)* You should be able to distinguish between the organic and aqueous layers within the vial. Gently, yet vigorously stir the contents of the vial (**~1 minute**) to mix the immiscible layers formed during the addition of the ether. ****NOTE:** *A small additional amount of diethyl ether may be necessary in order for the organic and aqueous layers to be distinguishable.*
5. Allow the contents of the vial to settle and the organic and aqueous layers to separate.
6. Remove the aqueous layer as detailed below *(which layer is it?)* and transfer it to a clean, dry 10 ml Erlenmeyer flask. ****NOTE:** *Never discard any layers during an extraction until all layers have been verified and the product has been successfully isolated.*
7. Add 1 mL of a 5% NaOH solution to the diethyl ether remaining in the vial, stir vigorously and allow the layers to settle.
8. Once again, remove the aqueous layer and add it to the first aqueous extract. Repeat the extraction one more time.
9. Using distilled water, wash the diethyl ether layer two times (**~1–2 mL**) and remove the aqueous layer as before.
10. Once the extraction process has been completed, the product will be isolated and the aqueous layers can be disposed of in the waste receptacle under the hood.
11. The diethyl ether layer containing the product must be dried to remove any excess water in the organic layer. The drying process is performed using sodium sulfate and the filtration column as shown in Figure 19(a). **NOTE:** *The filtration column* (**Pasteur pipette**) *should be* **LOOSELY** *packed with the cotton plug;* **DO NOT PACK THE COTTON FIRMLY!!**
12. Pre-weigh a 3 mL conical vial with a boiling chip and reserve to collect the effluent in it.
13. Run the diethyl ether solution through the micro filtration column and collect the effluent in the pre-weighed conical vial.
14. Rinse the column once with 1 mL of CH_2Cl_2. Collect this new effluent rinse in the pre-weighed 3 mL conical vial containing the initial effluent and boiling chip.
15. Carefully use a warm water bath to evaporate the solvent off.
16. Since the product, 1-ethoxy-4-ethyl benzene, has a very low melting point as a solid, the product will be a viscous liquid once all of the solvent has been evaporated off.

References
Organic Chemistry 7th ed.; John McMurry.
Introduction to Organic Laboratory Techniques: A Microscale Approach 3rd ed.; Pavia, Lampman, Kriz, and Engel.

Safety
1. **Take precaution when using the NaOH; it is a <u>CORROSIVE</u> reagent that can cause severe burns.**
2. **Take precaution when handling <u>HOT</u> objects.**
3. **Turn <u>OFF</u> any burners and/or hot plates when not in use.**
4. **Clean (using soap for glassware) and rinse (using distilled water for final rinse) <u>ALL</u> glassware and utensils <u>PRIOR</u> to returning them to your lab drawer.**

5. If exposure to <u>ANY</u> of the reagents occurs, be sure to wash hands and any other affected areas thoroughly with plenty of soap and water.

Disposal
Dispose of all excess organic liquid reagents in the "**Organic Liquid Wastes**" receptacle.
Dispose of the product in the "**Organic Liquid Wastes**" receptacle.
Dispose of the aqueous layer/washes in the "**Inorganic Liquid Wastes**" receptacle.
Dispose of any disposable pipettes and/or test tubes in the "**broken glass**" bin.

Characterization of the Product
1. Determine the experimental percent yield of your product.
2. Obtain an IR spectrum of your product using NaCl salt plates. Identify all functional and assign all relative vibrational peaks.

Postlab Questions
1. Why do you wash the diethyl ether layer containing your product with NaOH during the extraction procedure?
2. Outline the synthesis of each of the following ethers beginning with an alcohol and an alkyl halide or tosylate of the appropriate structure.

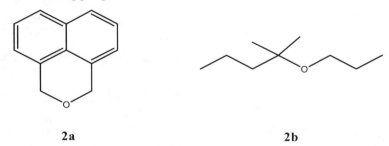

2a   2b

Sodium Borohydride Reduction of Ketone— Exp#16

Background
The reduction of a chemical compound is achieved by converting a molecule of a higher oxidation state to one of a lower oxidation state. This is accomplished by either adding hydrogen to the compound or by removing oxygen from the compound.

Theory
Aldehydes and ketones can be reduced, using the proper reagents, to form alcohols. There are two primary reagents that are utilized in the reduction of carbonyl compounds, lithium aluminum hydride ($LiAlH_4$) and sodium borohydride ($NaBH_4$). Although both of these reagents can reduce ketones and aldehydes, they differ in the ways listed below.

Terminology
- *Oxidation*: Loss of an electron or hydrogen; gain of oxygen.
- *Reduction*: Gain of an electron or hydrogen; loss of oxygen.

$NaBH_4$	$LiAlH_4$
Less reactive—mild reducing reagent	Very reactive—powerful reducing agent
Safer to handle	Must be handled carefully
Used in protic (e.g., alcohols) or aqueous solvents	Used in aprotic (e.g., diethyl ether) solvents
Reacts slowly with "neutral" H_2O and is stable in basic solutions	Reacts *violently* with H_2O and other hydroxylic solvents (produces H_2 gas)
Unreactive to the other functional groups listed for reactivity to lithium aluminum hydride	Reactive towards aldehydes, ketones, esters, epoxides, nitriles, and nitro groups

**The reduction of ketones can also be carried out via H_2 and metal catalyst; however, this method would also lead to the reduction of any alkenes present in the molecule.

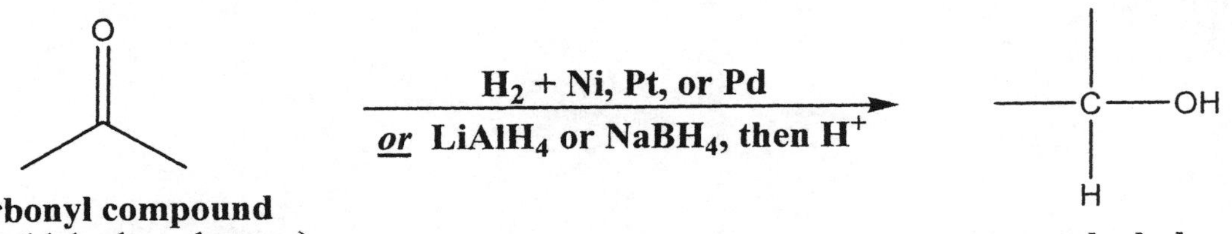

a carbonyl compound (e.g., aldehyde or ketone) $\xrightarrow{\text{$H_2$ + Ni, Pt, or Pd} \text{ or } \text{$LiAlH_4$ or $NaBH_4$, then H^+}}$ an alcohol

General Reaction

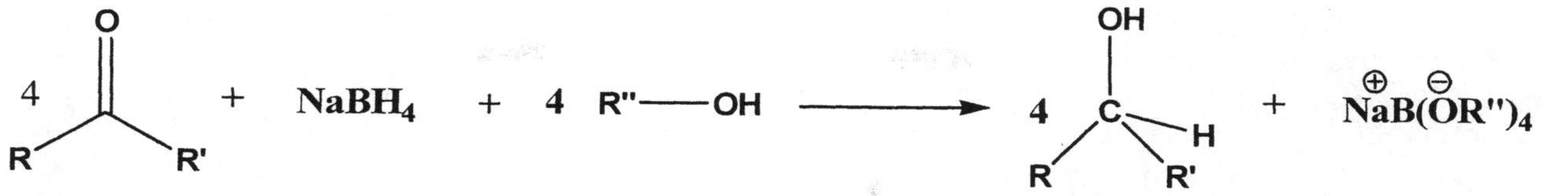

$$4 \; R\text{-CO-}R' + NaBH_4 + 4 \; R''\text{-OH} \longrightarrow 4 \; R\text{-CH(OH)-}R' + NaB(OR'')_4$$

General Mechanism

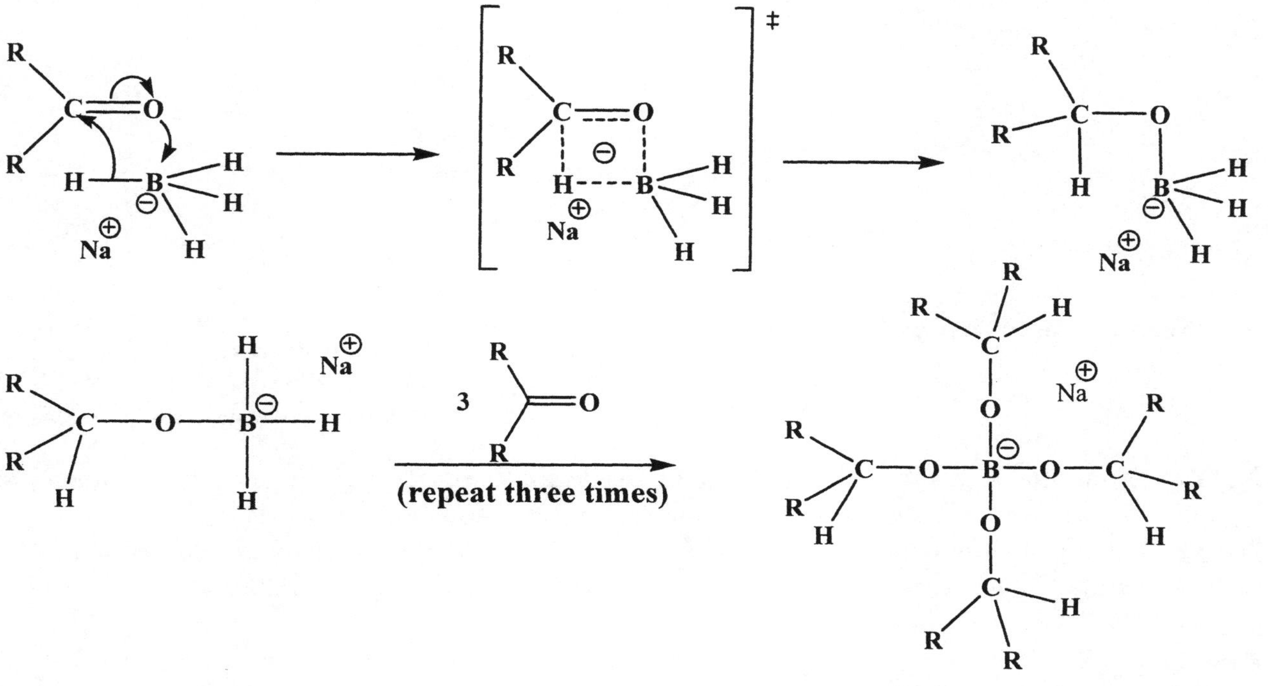

[Mechanism showing hydride transfer from BH_4^- to carbonyl via four-membered transition state, yielding alkoxyborohydride intermediate]

$$R_2CH\text{-O-}BH_3^- Na^+ \xrightarrow[\text{(repeat three times)}]{3 \; R_2C=O} \text{tetraalkoxyboron}$$

$$(R_2CHO)_4B^-Na^+ + 4 \; R'OH \longrightarrow 4 \; R_2CHOH + (R'O)_4B^-Na^+$$

Specific Reaction

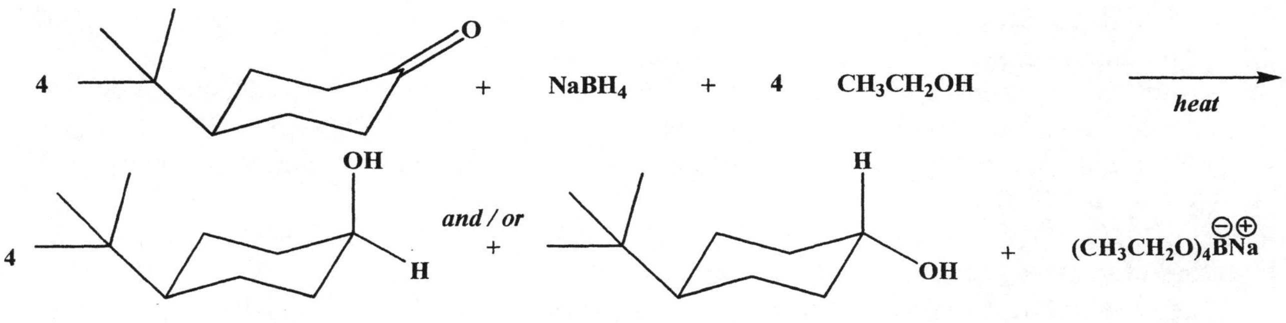

$$4 \; \text{(4-}t\text{-butylcyclohexanone)} + NaBH_4 + 4 \; CH_3CH_2OH \xrightarrow{heat}$$

4 (cis isomer, minor product) *and / or* (trans isomer, major product) + $(CH_3CH_2O)_4B^-Na^+$

4-*t*-butylcyclohexanol

Specific Mechanism

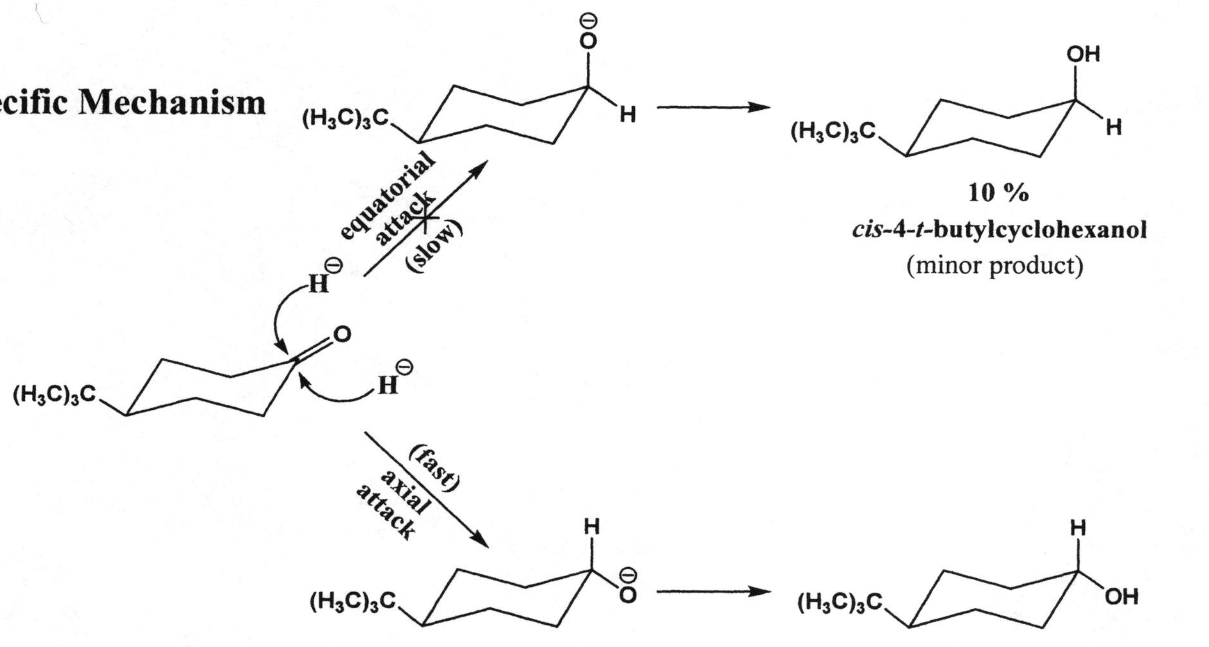

10 %
cis-4-t-butylcyclohexanol
(minor product)

90 %
trans-4-t-butylcyclohexanol
(major product)
thermodynamically more stable

Equatorial product (*trans*) is favored

NOTE: Axial attack of the carbonyl by the hydride ion is preferred; however, some equatorial attack will occur as well. This results in a mixture of the *cis*- and *trans*- isomers.

NOTE: A total of 4 moles of the $CH_3CH_2O\text{-}BH_3^-Na^+$ reacts with the carbonyl compound. The resultant product consists of 4 moles of the *trans*- isomer, or usually, a mixture of both.

Procedure

Reduction of 4-*t*-butylcyclohexanone:

1. Using a clean, dry 5 mL conical vial, weigh out 0.52 mmol of *t*-butylcyclohexanone. **NOTE: Use Tare method.**
2. Using the numerical scale on a 3 mL conical vial, measure 0.3 mL of ethanol and add it to the 5 mL conical vial from step 1.
3. Add a spin vane to the 5 mL conical vial and set to stir at room temperature until the entire solid has dissolved.
4. Measure 0.6 mL of $NaBH_4$ solution in ethanol using a 10 mL graduated cylinder and add it to the 5 mL conical vial. **NOTE: Be sure to use the <u>bottom</u> slurry solution, NOT the supernatant.**
5. Stir this mixture at room temperature for several minutes. A small amount of white precipitate should form during this time.
6. Continue to stir the solution for 15 more minutes—there may be additional lecture notes given at this time.

Isolation of the product, 4-*t*-butylcyclohexanol:

1. At the end of the stirring period, add a few drops of 3 M HCl. This is done to decompose the excess reducing agent. **CAUTION: *The excess $NaBH_4$ reacts to form H_2 gas—this may cause considerable foaming.***

Extraction of the product:
1. Once the foaming subsides, add 10 more drops of the 3 M HCl followed by 1.0 mL of distilled water and 0.5 mL of CH_2Cl_2, the extracting solvent (which layer is which ?).
2. Stir the mixture briskly for 1 minute. Cover the base, approximately 1 mm thickness, of a clean, dry 10 mL Erlenmeyer flask with anhydrous Na_2SO_4 (dehydrating agent). Let the two layers in the conical vial settle and with a filter-tipped Pasteur pipette, transfer the organic layer to the prepared Erlenmeyer flask.
3. Swirl the mixture in the flask several times, then let the solution stand for 5 minutes. **NOTE:** *This contains your first extract.*
4. Extract the aqueous layer remaining in the 5 mL vial from step 2 with another portion of 0.5 mL of *CH_2Cl_2* and stir the mixture briskly for ~1 minute. Let the two layers form and with a filter-tipped Pasteur pipette, transfer the organic layer to the Erlenmeyer flask (containing the first extract).
5. Extract the aqueous layer one more time with 0.5 mL *CH_2Cl_2* following the above procedure and transfer the organic layer to the Erlenmeyer flask (containing the first and second extractions). **NOTE:** *The dehydrating agent removes the solubilized water from the organic layer and forms a clump at the bottom of the Erlenmeyer flask. This step may need to be repeated; consult your lab instructor.*
6. Preweigh a 5 mL conical vial containing a boiling chip. Carefully decant the CH_2Cl_2 solution off from the $NaSO_4$ into the preweighed conical vial. Rinse the dehydrating reagent remaining in the flask with 0.5 mL of CH_2Cl_2 and transfer the rinse to the conical vial.
7. Evaporate the CH_2Cl_2 in a warm water bath (40 °C) under the hood using a hot plate. **NOTE:** *The vial may have to be lifted up and down to prevent the solution from overflowing during this process.* Continue the evaporation process until bubbles are no longer visible.
8. After the volume of solvent has noticeably decreased, consult your instructor. Once the instructor has approved your experiment thus far, allow the vial to cool to room temperature. Once the vial has cooled completely, weigh the vial. **NOTE: The vial should be reheated, cooled, and reweighed until a constant weight is obtained.**
9. As the vial cools, your product should crystallize. If it does not, place it in an ice bath and scratch the side of the vial using a spatula until your product comes out.
10. Allow the product to stand at room temperature for 15 minutes to dry.

Characterization of the product, 4-*t*-butylcyclohexanol
1. Run an IR spectrum on your product using KBr pellet. Assign the major peaks.
2. Obtain the melting point of the isolated product and compare it to the known melting point found in literature.
3. Determine the limiting reagent, excess reagent, theoretical yield, and percent yield. **NOTE:** *Show all work and setup for equations.*

References
Organic Chemistry 7th Edition; John McMurry.
Introduction to Organic Laboratory Techniques: A Microscale Approach 3rd ed.; Pavia, Lampman, Kriz, and Engel.

Safety
1. Take precaution when handling <u>3M HCl. If it comes in contact with skin, rinse the affected area with copious amount of water.</u>
2. CH$_2$Cl$_2$ is carcinogen, so if any exposure to this reagent occurs, be sure to wash the affected area with ample amount of water.
3. Turn **OFF** any burners and/or hot plates when not in use.
4. Clean (using soap for glassware) and rinse (using distilled water for final rinse) **ALL** glassware and utensils **PRIOR** to returning them to your lab drawer.

Disposal
Dispose of the liquid reagents in the "Organic Liquid Wastes" receptacle.
Dispose of the solid product in the "Organic Solid Wastes" receptacle.
Dispose of any disposable pipettes and / or test tubes in the "broken glass" bin.
Dispose of excess CH$_2$Cl$_2$ in "Halogenated Organic Wastes."

Postlab Questions
1. Draw the products for the following reduction reactions.

(a) [structure: aromatic ether with HO and OHC substituents on one ring, and H$_2$C-C(=O)-O-CH$_3$ on the other ring]

1. NaBH$_4$, H$_2$O, OH$^{\ominus}$
2. H$_2$O, H$^{\oplus}$

(b) [structure: H$_3$C-O-C(=O)-CH$_2$- attached to a bicyclic ketone with CH$_3$ substituent]

1. LiAlH$_4$, THF
2. H$_2$O, H$^{\oplus}$

2. NaBH$_4$ is a less powerful reducing reagent compared to LiAlH$_4$. In what ways do these two reagents differ?

Setup for Grignard Reaction Lab
1. You **MUST** clean, rinse with distilled water, then rinse with **ACETONE** the following pieces of equipment: **10 mL round-bottom flask, Claisen adapter, condenser, drying tube, (2) 3 mL vials, 5 Pasteur pipettes, spin vane, and a metal spatula** <u>(*NOT* your scupula!)</u>
2. Following the acetone rinse, wrap these items in an adequate piece of aluminum foil, label it for your lab group, and place it in the drying oven as per your instructor.

Grignard Reaction—Exp#17

Background

Organometallic reagents are very important in organic synthesis. An organometallic compound is a compound that contains a carbon-metal bond. Unlike carbon-oxygen bonds (alcohols or ethers), carbon-nitrogen bonds (amines), or carbon-halogen bonds (alkyl halides), in which carbon is bonded to a *more* electronegative atom, the carbon is bonded to a *less* electronegative atom in organometallic compounds.

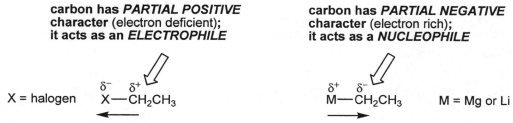

Formation of the Grignard Reagent

Grignard reagents are prepared by the reaction of magnesium metal with an alkyl halide, vinyl halide, or an aryl halide (i.e. aryl chloride) in an ether solvent such as THF or diethyl ether.

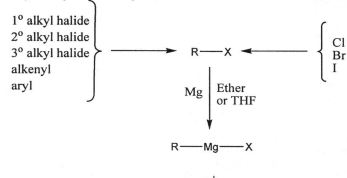

For example,

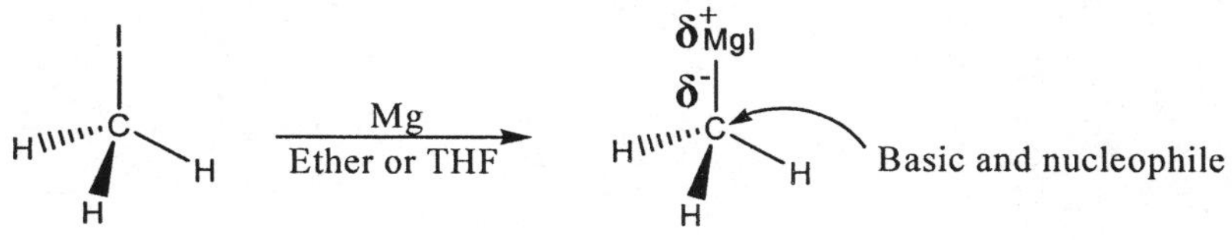

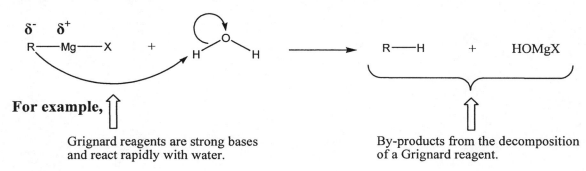

Grignard Reagent is Very Sensitive to Presence of Water:

$\overset{\delta^-}{R}-\overset{\delta^+}{Mg}-X$ + H–O–H ⟶ R–H + HOMgX

For example, ⇧

Grignard reagents are strong bases and react rapidly with water.

By-products from the decomposition of a Grignard reagent.

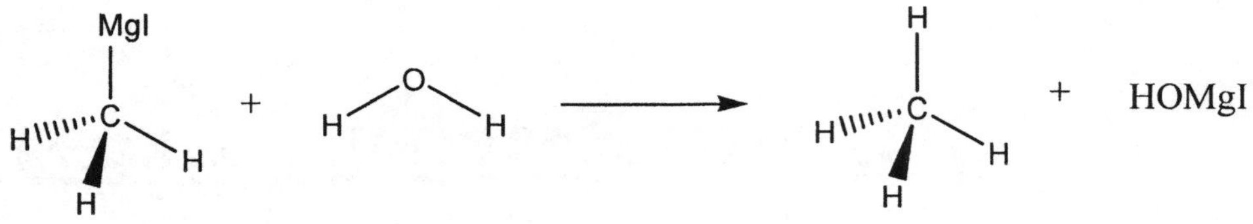

Methylmagnesium iodide + H₂O → Methane + HOMgI

Theory

The Grignard reaction is a distinct organometallic chemical reaction between either an alkyl- or aryl-magnesium halide **(Grignard reagent)** and an electrophile **(such as a carbonyl compound)**. The Grignard reagent acts as a strong base due to the direct connectivity of the electron-rich carbon atom to the magnesium atom. This causes the Grignard reagents to react quickly with most acids including water, alcohol, amines, and thiols. The most common organometallic reagents are organolithium and organomagnesium compounds. Grignard reagents are organomagnesium compounds named after their discoverer, Francois Auguste Victor Grignard (Nobel Prize winner in chemistry, 1912).

General Reaction

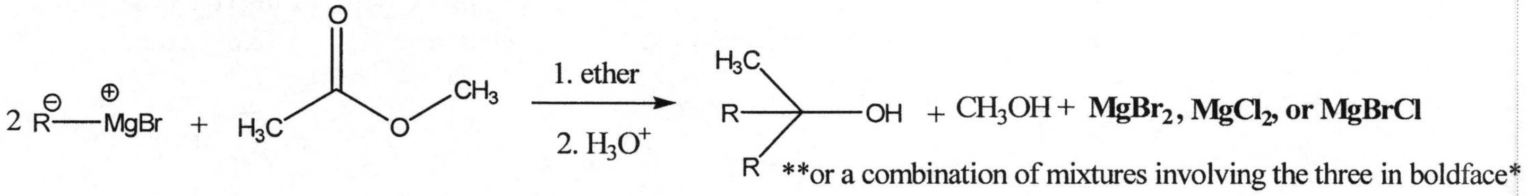

$2\ R^{\ominus}\text{—MgBr}^{\oplus} + H_3C\text{—C(=O)—O—CH}_3 \xrightarrow[\text{2. }H_3O^+]{\text{1. ether}}$ R₂C(CH₃)(OH) + CH₃OH + **MgBr₂, MgCl₂, or MgBrCl**

**or a combination of mixtures involving the three in boldface*

Specific Reaction

Step 1:

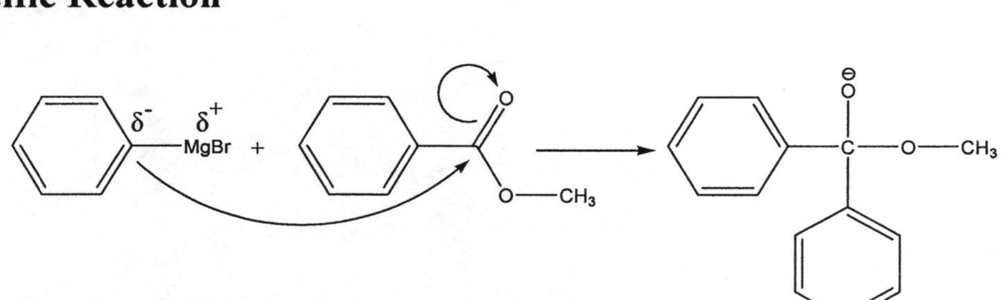

Ph—MgBr (δ⁻, δ⁺) + PhC(=O)—O—CH₃ → Ph₂C(O⁻)—O—CH₃

Step 2:

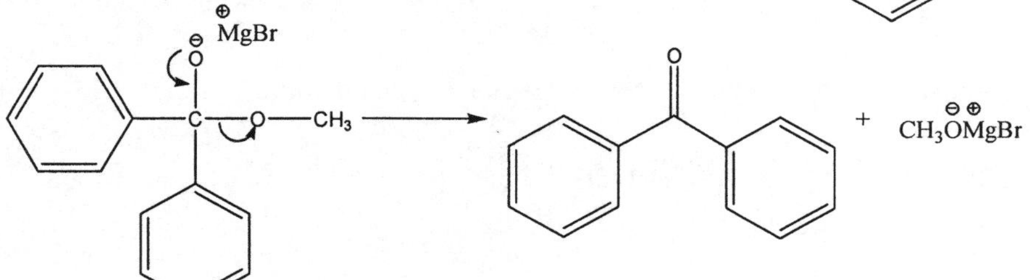

Ph₂C(O⁻MgBr⁺)—O—CH₃ → Ph₂C=O (benzophenone) + CH₃O⁻MgBr⁺

Step 3:

Ph—MgBr (δ⁻, δ⁺) + Ph₂C=O → Ph₃C—O⁻ MgBr⁺

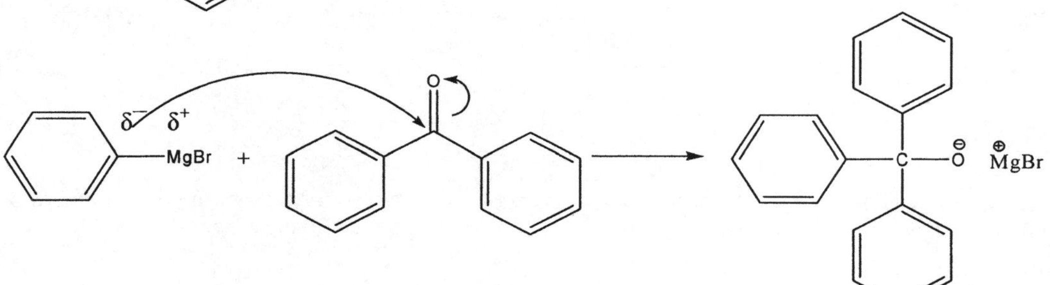

Step 4:

[Reaction scheme: triphenylmethoxide magnesium bromide + CH₃OMgBr reacts with HCl/H₂O to give triphenylmethanol (Ph₃C–OH) + CH₃OH + MgBr₂, MgCl₂, or MgBrCl]

Procedure

or a combination of mixtures involving the three in boldface

1. Before beginning this experiment, ensure that <u>**ALL**</u> of the glassware involved is free of water prior to preparing the Grignard reagent.
2. Remove the glassware from the oven and unwrap the aluminum foil.
3. Obtain one needle and ***three*** 1.0 mL syringes or dry, calibrated Pasteur pipettes per group from your instructor:
   - One for the bromobenzene (vial #1)
   - One for the methyl benzoate (vial #2)
   - One for adding ether to the reaction—*this syringe will be used throughout the experiment for the introduction of diethyl ether into the reaction solution.* <u>**PLEASE**</u> *keep it safe, clean, and dry when not in use.*

Formation of phenyl magnesium bromide:

1. Assemble the reaction apparatus as displayed below.
2. **DURING THIS EXPERIMENTAL PROCEDURE, REMEMBER TO RECAP <u>ALL</u> REAGENTS AFTER USE TO KEEP THEM FREE FROM WATER VAPOR CONTAMINATION!!**

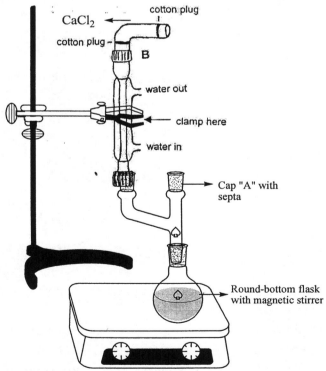

Figure 21: Apparatus for Grignard reaction

3. Obtain two clean, dry 3 mL conical vials. Label the vials "vial #1" and "vial #2."
4. To vial #1 add:
   - 2.5 mmol bromobenzene
   - 1.5 mL diethyl ether
   - ***Immediately*** use a cap and septum to seal the mixture, then gently shake to mix.
5. To vial #2 add:
   - 1.00 mmol methyl benzoate
   - 1.5 mL diethyl ether
   - <u>Immediately</u> use a cap and septum to seal the mixture, then gently shake to mix.
6. To a 10 mL round-bottom flask equipped with a spin vane, add 2.1 mmol magnesium turnings and 1.0 mL of diethyl ether.
7. Using the opening through "cap A" and a spatula, add one small piece of iodine crystal. Immediately reseal the apparatus with your septum and cap. ****NOTE:** *The iodine crystal aids in initiating the radical coupling reaction for the Grignard reagent formation.*
8. Follow Figure 22 for transferring the contents of the reagent vials to the round-bottom flask reaction apparatus.

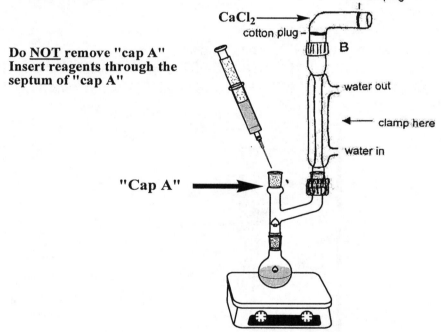

Figure 22: Grignard setup

9. Use a 1.0 mL syringe to add 1.0 mL of diethyl ether through the septum "cap A" to rinse the iodine down into the flask.
10. Set the hot plate temperature to 2–3. Reflux the solution in the round-bottom flask while stirring for approximately 5 minutes.
11. Transfer 4–5 drops of the reagent mixture from vial #1 through the septum of "cap A" using one of your 1.0 mL syringes. Continue to reflux for 3 more minutes. ****NOTE:**

There should be a color change in the solution from a dark brown to nearly colorless or light brown. If there is no visible color change, notify your instructor immediately.

12. Following the 3 minutes of reflux, transfer the rest of the contents of vial #1 to the round-bottom flask *over a 10-minute period*.
13. Continue to reflux for approximately 10–15 minutes or until nearly all of the magnesium has reacted (*a minimal visible amount remaining*) and solution has turned a clear amber color.

Formation of triphenylmethanol—the product:
1. Using a second clean syringe, and *over a 2-minute period*, transfer all of the reagent contents of vial #2 through the septum of "cap A" into the round-bottom flask. Continue refluxing during this process. ****NOTE:** *There may be a slight change in color of the solution.*
2. Continue refluxing the solution for another 30 minutes. ****NOTE:** *If the solvent level drops too low during this portion of the reflux process, add more diethyl ether through the septum of "cap A" using the same syringe as before.*
3. Once the reflux is complete, there should be a visible formation of the alkoxide salt as it precipitates out of the solution.
4. While still capped and sealed, allow the solution to cool to room temperature. Use an ice water bath to cool the round-bottom flask. Use as small a beaker as you can comfortably fit the round-bottom flask into to prevent accidental spillage.
5. Take your round-bottom flask into the fume hood and carefully remove "cap A."
6. Add 20 drops of distilled H_2O dropwise. Then, while swirling the solution, add drops of a 3 M HCl solution to the reaction. It may also be necessary to add more diethyl ether to the solution as well.
7. Once all of the solids are dissolved and the H_2 gas evolution has ceased, a clear aqueous layer should be visible. ****NOTE:** *It is normal to observe H_2 gas formation as bubbles. This is a result of a reaction between the excess Mg present, HCl and H_2O..*
8. Once the H_2 gas has completely evolved, remove the reaction mixture from the ice water bath and the reaction apparatus may be disassembled.
9. A biphasic clear mixture should be visible with the organic diethyl ether layer at the top and an acidic aqueous layer at the bottom. ****NOTE:** *If a clear biphasic mixture does not occur, consult your lab instructor immediately.*

Isolation and characterization of the product—Triphenylmethanol via extraction:
1. Use a filter-tip pipette to transfer the biphasic mixture from the round-bottom flask to a clean, dry 5 mL conical vial.
2. Use a small amount of diethyl ether to rinse the round-bottom flask following the transfer. Transfer the rinse solution to the same 5 mL conical vial. ****NOTE:** *If the solution exceeds 5 mL, divide the solution between two 5 mL flasks and continue the procedure as written.*

3. Using forceps or tweezers, remove the spin vane from the round-bottom flask and into the 5 mL vial being careful to avoid excessive splashing or loss of solution. Gently stir the contents of the vial for ~1 minute.
4. While using a micropipette, carefully remove the aqueous layer (which layer?) of the solution and transfer it to a clean, dry 50 mL Erlenmeyer flask and set aside. Avoid accidental transfer of the product in the organic layer.
5. Add 1–2 mL of distilled water to the diethyl ether organic solution remaining in the 5 mL conical vial. Allow the solution to stir for ~1 minute.
6. Remove this second aqueous layer using a pipette—it may be necessary to repeat this rinsing process once more. Include these rinses with the initial rinse within the 50 mL Erlenmeyer flask.
7. As a final rinse, add 1 mL of a saturated NaCl solution to the 5 mL vial containing the diethyl ether organic layer. Stir this solution vigorously and remove the aqueous layer. Add this rinse to the same 50 mL flask.

Drying of the crude product:
1. Cover the base of a second clean, dry 10 mL Erlenmeyer flask with anhydrous Na_2SO_4 **(~1 mm in thickness)**. Transfer the diethyl ether solution containing the product to this flask. Swirl the flask several times and then let it stand for ~5 minutes. ****NOTE:** *You may have to repeat this step—consult your lab instructor.*
2. Preweigh a clean, dry Craig tube. Carefully using a filter-tip pipette transfer diethyl ether solution into this preweighed Craig tube containing a boiling chip. Use ~0.5 mL of diethyl ether to rinse the Erlenmeyer flask, then use a Pasteur pipette to transfer this rinse to the Craig tube. ****NOTE: It may be necessary to divide the diethyl ether solution between two Craig tubes if the volume exceeds 2mL.**
3. Use a warm water bath (~35 °C) to evaporate most (**NOT all**) of the diethyl ether off from the Craig tube. Remove the Craig tube from the water bath and add 20 drops of hexane to the remaining solution. ****NOTE:** *A precipitate should form; if not, consult your instructor immediately.*
4. Redissolve the precipitate by warming the solution in the warm water bath. Allow the solution to cool *slowly* to room temperature.
5. Once completely cooled, place the Craig tube in an ice water bath.
6. Wrap or tie a piece of wire around your Teflon plug. Use this plug to seal your Craig tube. Place the Craig tube/Teflon plug in a centrifuge tube as per your lab instructor.
7. Isolate the product via a **2 minute** centrifugation. ****NOTE:** *Be sure to balance the centrifuge with another group's sample of equal quantity or with an other centrifuge tube filled with enough water to balance the centrifuge.*
8. Preweigh a clean, dry watch glass and record its mass in your lab notebook. Use this watch glass to collect your product from the Craig tube. Record the weight of the product in your lab notebook.

Characterization of the Product
1. Obtain the experimental melting point value of your product using a capillary tube. Record the known melting point of your product in your lab notebook along with the experimentally determined value.
2. Obtain an IR spectrum of your product using KBr pellet. Identify all functional groups and assign all relative vibrational peaks.
3. Calculate the experimental percent yield of the product.

Reference
Organic Chemistry 7th ed.; John McMurry.
Introduction to Organic Laboratory Techniques: A Microscale Approach 3rd ed.; Pavia, Lampman, Kriz, and Engel.

Safety
1. **Take precaution when handling the HCl because it is <u>CORROSIVE</u>. If there is contact with the skin; rinse the affected area with copious amounts of water.**
2. **Take precaution when handling the diethyl ether because it is <u>FLAMMABLE</u>. Do <u>NOT</u> expose this reagent to <u>ANY</u> heat source or open flame.**
3. **Take precaution when handling <u>HOT</u> objects.**
4. **Turn <u>OFF</u> any burners and/or hot plates when not in use.**
5. **Clean (using soap for glassware) and rinse (distilled water for final rinse) <u>ALL</u> glassware and utensils <u>PRIOR</u> to returning them to your lab drawer.**
6. **If exposure to <u>ANY</u> of the reagents occurs, be sure to wash hands and any other affected areas thoroughly with plenty of soap and water.**

Disposal
Dispose of the liquid bromobenzene in the "Halogenated Organic Liquid Wastes" receptacle.
Dispose of all other liquid reagents in the "Organic Liquid Wastes" receptacle.
Dispose of the solid product in the "Organic Solid Wastes" receptacle.
Dispose of any disposable pipettes and/or test tubes in the "broken glass" bin.
Dispose of **any syringes or needles** in the **RED BOX** under the hood.

Postlab Questions:
1. Provide the reagents necessary for preparing the following compounds by the Grignard reaction.

2. During the extraction of your product, what is the role of HCl in the formation of the triphenylmethanol? Draw the complete reaction using arrow formalism.

Fischer Esterification—Exp#18

Background
Fischer esterification is used to prepare esters via a condensation of a carboxylic acid and an alcohol. A strong acid is used in this reaction as a catalyst for the condensation to take place.

Fischer esterification has two specific concepts to keep in mind:

- **The reaction is acid-catalyzed:** A minimal amount of a strong acid is needed to "activate" the carbonyl carbon of the carboxylic acid toward a nucleophilic attack by the alcohol. Without the acid catalyst, the reaction would occur, yet at a much slower rate.
- **The reaction is reversible:** In a reversible reaction, the reaction eventually reaches equilibrium. Based upon Le Chatlier's Principle, a disturbance in this equilibrium will cause the reaction to shift accordingly to offset the disturbance. Therefore, a reversible reaction can be driven *towards* the desired product by either using an excess of one of the reagents or removal of the product as it is being produced.

Theory
In this experiment you will prepare an ester "isopentyl acetate" via Fischer esterification. This ester is often referred to as banana oil, because it has the familiar odor of this fruit.

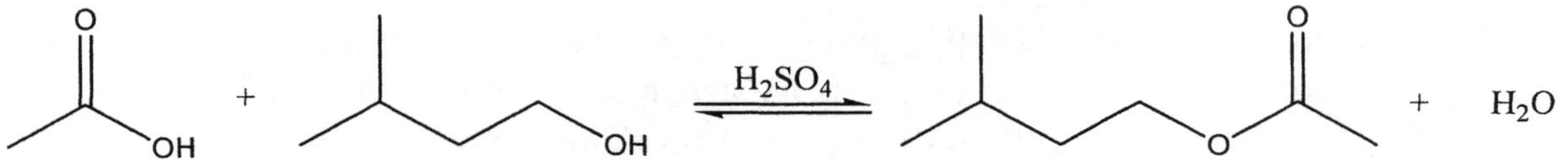

$$\text{acetic acid} + \text{isopentyl alcohol} \xrightleftharpoons{H_2SO_4} \text{isopentyl acetate} + H_2O$$

Isopentyl acetate is prepared by the direct esterification of acetic acid with isopentyl alcohol. Since this reaction like other organic synthesis is reversible, we should not expect a high yield for the above reaction (yield ~ 60–70 %). This percentage yield can be increased to nearly 100% by using an excess amount of one starting material such as alcohol or acid and by removing the water that forms during the reaction. In these ways, the equilibrium can be shifted in favor of ester formation. In this experiment acetic acid is used in excess because it is less expensive than isopentyl alcohol and more easily removed from the reaction mixture by extraction method.

In the isolation step, much of the excess acetic acid is removed by extraction with sodium bicarbonate and water. After drying with anhydrous sodium sulfate, the ester is purified by distillation. The purity of the liquid product is analyzed by performing boiling point determination and IR spectroscopy.

General Reaction

Basic:

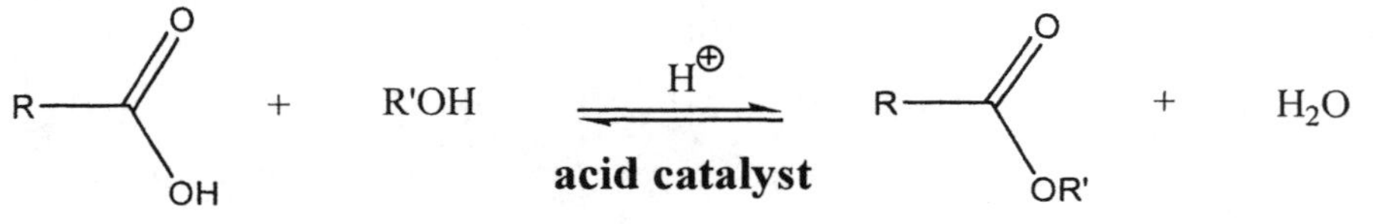

$$R\text{-COOH} + R'OH \xrightleftharpoons{H^{\oplus} \text{ acid catalyst}} R\text{-COOR'} + H_2O$$

carboxylic acid   alcohol   ester   water

Detailed:

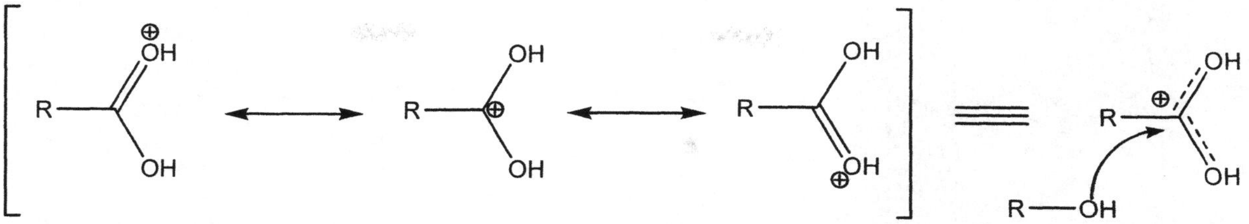

[resonance structures for protonated carboxylic acid]

$$R-COOH + H_2SO_4 \rightleftharpoons R-C(OH)(OH)^+ + HSO_4^-$$
acid catalyst

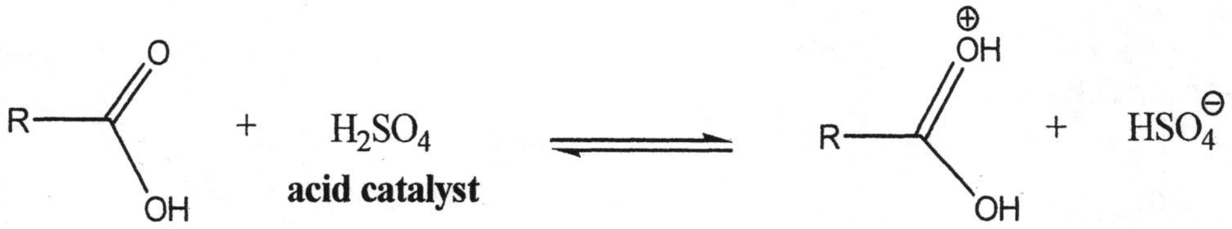

General Methods for Esterification
Method A:

$$R-COOH + R'OH \underset{}{\overset{H^+}{\rightleftharpoons}} R-COOR' + H_2O$$

Method B:

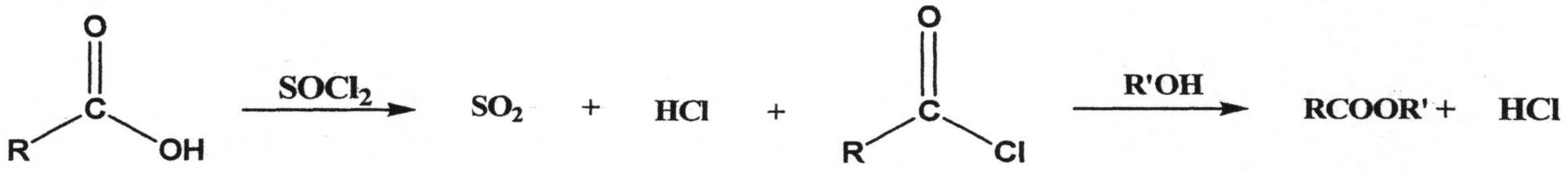

$$R-COOH \xrightarrow{SOCl_2} SO_2 + HCl + R-COCl \xrightarrow{R'OH} RCOOR' + HCl$$

Specific Mechanism
Step 1: Protonation of the carboxylic acid by the acid catalyst.

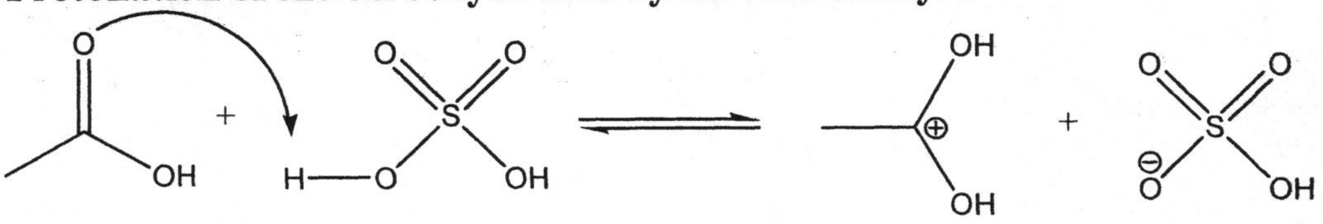

Step 2: Nucleophilic attack of the protonated carboxylic acid by the alcohol.

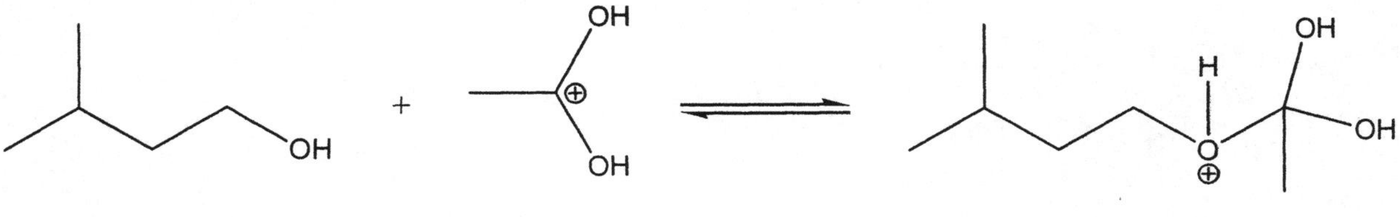

Step 3: Intramolecular proton transfer to form an oxonium ion, followed by loss of water.

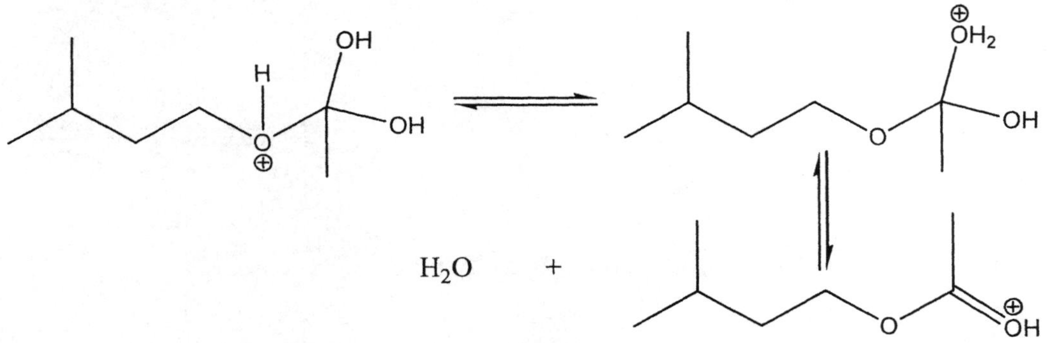

Step 4: Deprotonation by the carboxylic acid or by bisulfate ion to form the ester and more protonated acid to drive the reaction.

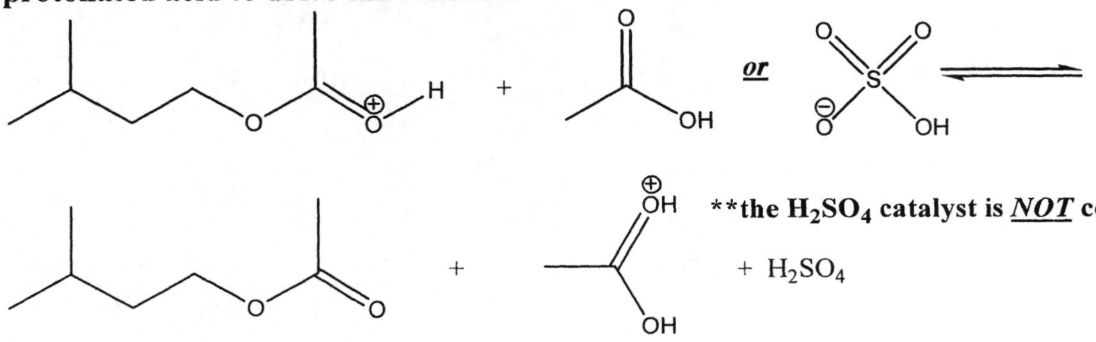

the H$_2$SO$_4$ catalyst is *NOT* consumed

+ H$_2$SO$_4$

Special Instructions
Because 45 minutes of reflux is required, you should start the experiment at the very beginning of the laboratory period. During the reflux period, your instructor may lecture.

> **NOTE:** The reagents and product in this experiment all have strong odors. To minimize odors in the laboratory environment, prepare a drying tube with $CaCl_2$ as demonstrated by the instructor.

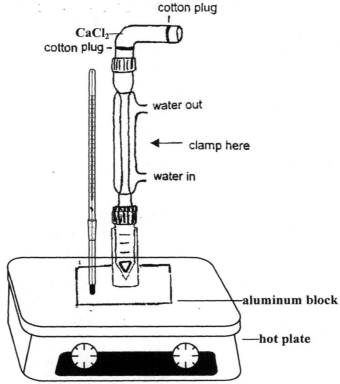

Figure 23: Apparatus setup for experiment

Procedure
Formation of isopentyl acetate:
1. Prepare a drying tube with $CaCl_2$ as demonstrated by the instructor and place it on top of the air condenser—refer to Figure 23. This is used to help minimize the odors formed during the reaction procedure.
2. Place 7.3 mmol of isopentyl alcohol and 28 mmol of acetic acid *(4 fold excess)* into a 5 mL conical vial with a spin vane.
3. Under a fume hood, add 5 drops of concentrated H_2SO_4 to your reaction mixture.
4. Fit the conical vial with the air condenser/drying tube and proceed to heat the reaction mixture to reflux while stirring (heat to ~140–150 °C).
5. Once the reaction mixture begins to reflux, continue heating for 45 minutes. Record any observations in your lab notebook.
6. Following the reflux, remove the reaction vial from the hot plate with the condenser/drying tube still attached. Allow the reaction vessel to cool to room temperature.
7. Once the vial has cooled, cool it further by placing it in an ice water bath.

Isolation of the product:
1. Once the reaction mixture has sufficiently cooled, disconnect the condenser/drying tube. Add *cold* water to the vial until it is almost full.
2. Stir the mixture rapidly with the spin vane for about 1 minute. ****CAUTION:** *Avoid splashing of the solution.*
3. Discontinue stirring and allow the organic and aqueous layers to separate. ****NOTE:** *The organic ester layer should be floating on top of the aqueous layer.*
4. Using a Pasteur pipette, carefully remove the aqueous layer and transfer it to a clean, dry 50 mL beaker. Add 1.5 mL of diethyl ether to the organic layer remaining in the conical vial. ****NOTE:** *The diethyl ether is added to extract the ester product from the organic layer (what is the composition of the organic layer?).*
5. Once again, fill the conical vial with cold water. Stir the solution, remove the aqueous layer and transfer it to the same 50 mL beaker as before. ****NOTE:** *Additional diethyl ether may be added if there is difficulty in identifying the two layers.*
6. Add 1 mL of a 5 % $NaHCO_3$ solution, stir and discard the aqueous layer. Record any observations in your laboratory notebook.
7. In the same manner as before, wash the diethyl ether layer with 1 mL of saturated aqueous NaCl.
8. Remove as much of the aqueous layer as possible and add to the other aqueous washes.
9. Cover the base of a 25 mL Erlenmeyer flask with ~1 mm thickness of anhydrous Na_2SO_4.
10. Transfer the diethyl ether solution containing the product to the prepared Erlenmeyer flask. Swirl the flask several times and let it stand for 5 minutes. ****NOTE:** *This step may be repeated if the mixture clumps and doesn't flow or looks cloudy with obvious drops of water. Repeat by transferring the ester to a new clean, dry 25 mL Erlenmeyer flask and add a NEW 0.5 g portion of Na_2SO_4 to complete the drying.*
11. Carefully decant the diethyl ether solution into a clean, dry 3 mL conical vial containing a spin vane.
12. Rinse the salt with 1 mL of ether and swirl several times. Transfer the rinse to the same 3 mL conical vial. ****NOTE:** *The volume of the solution should be at least 2 mL in order to have a successful distillation during the next step.*

Purification of the product:
1. Attach a Hickman still with a thermometer adapter to your conical vial.
2. A thermometer is placed in the Hickman still and adjusted so that the thermometer bulb is *below* the lip of the Hickman still. To speed up distillation, insulate your system by wrapping the Hickman still and the vial snugly, yet not tightly, with aluminum foil. Also, make sure your system is *NOT* sealed.
3. Adjust the hot plate to ~3. Begin heating the reaction vessel. ****NOTE:** Be sure to note which component will boil off first, second, and third.
4. Using a Pasteur pipette to remove the first fraction at about 35 °C. Remove the second fraction at about 100 °C.
5. Carefully, and slowly continue to raise the temperature until the ester starts to distil. ****NOTE: Be sure to note the temperature of the distillation and record it in your lab notebook, when you collect each fraction.**
6. When only a few drops of liquid remain in the vial, turn off the hot plate. Preweigh a third conical vial.
7. Using a Pasteur pipette, transfer the **distillate** *(ester contained in third fraction)* to the preweighed conical vial. Record the mass of the product in your lab notebook.

8. Calculate the theoretical, actual, and percent yields. Show all calculations and setup.

Characterization of the Product
1. Record the known boiling point of your product in your lab notebook along with the experimentally determined value.
2. Obtain an IR spectrum of your product using NaCl salt plates. Identify all functional groups and assign all relative vibrational peaks.
3. The ^{1}H NMR spectrum of isopentyl acetate will be provided for you. Assign all peaks in the spectrum.
4. Calculate the experimental percent yield of the product.

References
Organic Chemistry 7th ed.; John McMurry.
Introduction to Organic Laboratory Techniques: A Microscale Approach 3rd ed.; Pavia, Lampman, Kriz, and Engel.

Safety
1. **Be careful when dispensing sulfuric and glacial acetic acids. They are very <u>CORROSIVE</u> and will attack your skin if you make contact with them. If you get one of these acids on your skin, wash the affected area with copious quantities of running water for 10–15 minutes.**
2. **Take precaution when handling <u>HOT</u> objects.**
3. **Turn <u>OFF</u> any burners and/or hot plates when not in use.**
4. **Clean (using soap for glassware) and rinse (using distilled water for final rinse) <u>ALL</u> glassware and utensils <u>PRIOR</u> to returning them to your lab drawer.**
5. **If exposure to <u>ANY</u> of the reagents occurs, be sure to wash hands and any other affected areas thoroughly with plenty of soap and water.**

Disposal
Dispose of any excess reagents in the "Organic Liquid Wastes" receptacle.
Dispose of the distillates and product in the "Organic Liquid Wastes" receptacle.
Dispose of any disposable pipettes and/or test tubes in the "broken glass" bin.

Postlab Questions
1. Write chemical equations to show what happens when NaHCO$_3$ is added during the extraction step of the experiment.
2. Fischer esterification does not work well with most tertiary alcohols. What would the major product of the reaction be if *t*-butyl alcohol was combined with acetic acid in the presence of H$_2$SO$_4$?
3. Many esters have characteristic odors (usually pleasant). Methyl salicylate exhibits a pleasant wintergreen odor. This ester can be prepared from salicyclic acid and methanol via a Fischer esterification. Which one of these reagents would you use in excess in order to drive the reaction to product? Explain how you would isolate the ester product from the reaction mixture.

salicylic acid + CH$_3$OH $\xrightleftharpoons{H_2SO_4}$ methyl salicylate + H$_2$O

Common Esters and Their Respective Fragrances

Ester	Fragrance
Isoamyl acetate	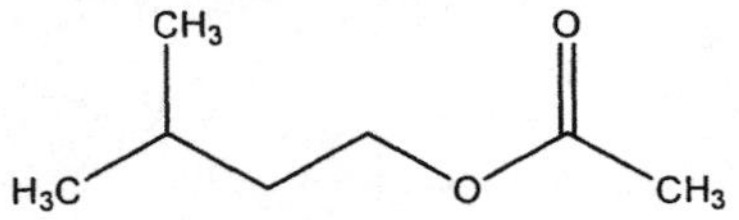 banana
Benzyl acetate	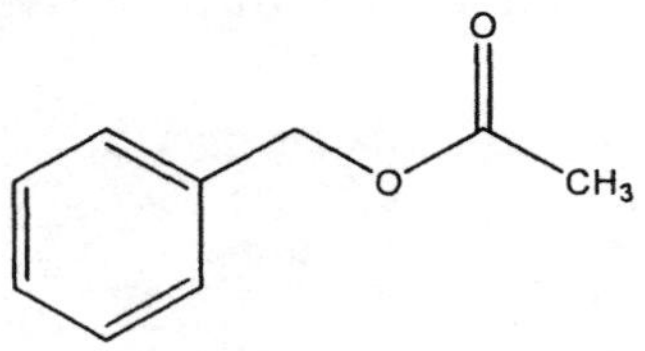 peach
Methyl butyrate	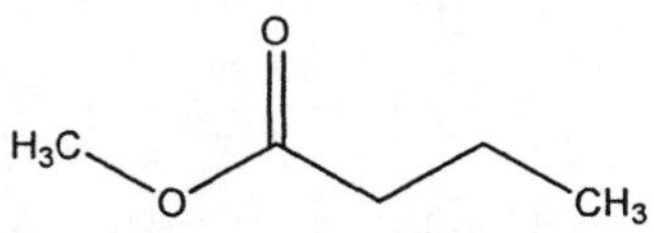 apple
Ethyl butyrate	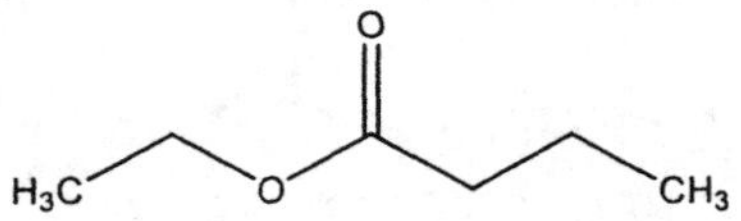 pineapple
Isobutyl propionate	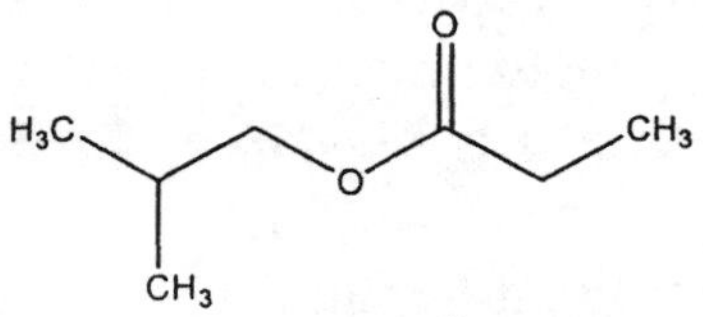 rum
Benzyl butyrate	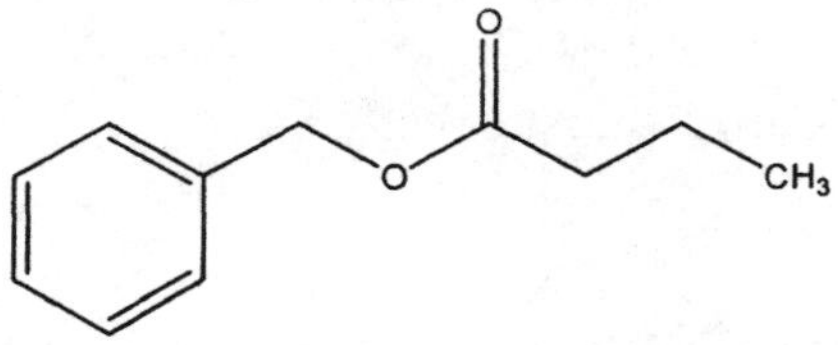 cherry
Methyl anthranilate	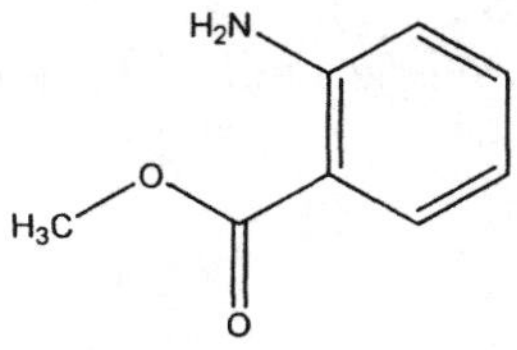 grape
Ethyl phenylacetate	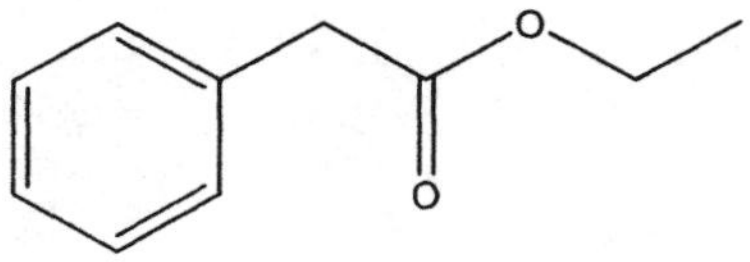 honey
Methyl salicylate	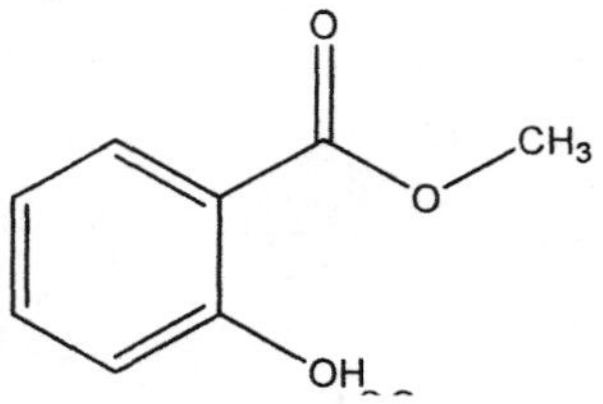 wintergreen

Dye-Coupling and Diazo-Imaging—Exp#19

Background
The practice of using dyes is an ancient art. There is substantial evidence that plant dyestuff were known long before humans began to keep written history. Before the 20th century, practically all dyes were obtained from natural plant or animal sources; dyeing was a complicated and secret art passed from one generation to the next. Dyes were extracted from plants macerating the roots, leaves, or berries in water. The extract was often boiled and then strained before use. In some cases, it was necessary to make the extraction mixture acidic or basic before the dye could be liberated from the plant tissues. Applying the dyes to cloth was also a complicated process.

The followings are some examples of plant dyestuffs: alizarine, henna, indicant, indoxyl, etc.

After 1856, synthetic dyes were started to be developed among these dyes. The AZO dyes were one of the most common types of dye still in use today. They are used as dyes for clothing, as food dyes and as pigments in paints. In addition, they are used in printing inks and in certain color printing processes. Azo dyes have the basic structure:

Ar - N = N - Ar'
Ar' = activated aromatic compound

Some examples of azo dye applications:
Methyl orange is prepared by the diazo coupling reaction. It is prepared from sulfanilic acid and N, N-dimethylaniline. The first product obtained from the coupling is the bright red acid form of methyl orange, called helianthin. In base, helianthin is converted to the orange sodium salt, called methyl orange.

"azo" functional group

Methyl orange: Used as a pH indicator

FD & C Red No. 40: An azo dye approved as an additive to food, drugs, and cosmetics by the FDA.
NOTE: The anionic sulfonate groups attached are soluble in water, a necessity for some of the dyeing applications.

Solvent Yellow 14: A type of neutral azo dye that is *NOT* soluble in water, yet can still be used as a pigment (such as for paints) and in the dying of fabrics.

Theory
Several azo dyes can be prepared using *diazo-imaging* and processing. The *diazo-imaging* technique is a less-sophisticated version of modern photocopying. The synethesis of the azo dyes in this experiment are produced in two major phases: *diazotization* and *coupling*.

General Diazotization Mechanism

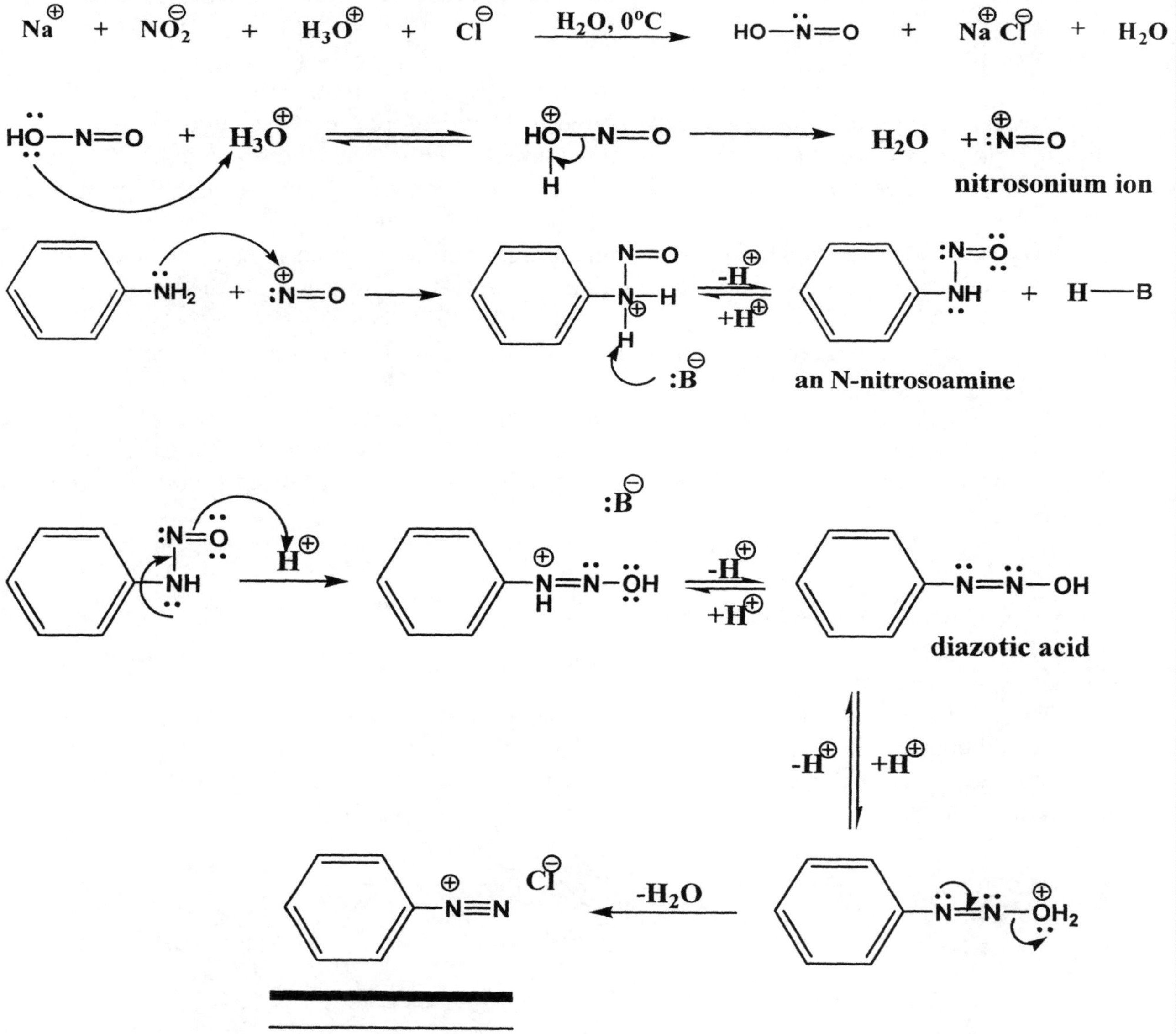

[Reaction: Na⊕ + NO₂⊖ + H₃O⊕ + Cl⊖ →(H₂O, 0°C) HO–N=O + Na⊕Cl⊖ + H₂O]

[Reaction: HO–N=O + H₃O⊕ ⇌ HO⊕(H)–N=O → H₂O + :N=O⊕ (nitrosonium ion)]

[Reaction: PhNH₂ + :N=O⊕ → Ph–N⊕H(N=O)–H ⇌ (−H⊕/+H⊕) Ph–NH–N=O + H–B (an N-nitrosoamine)]

[Reaction: Ph–NH–N=O + H⊕ → Ph–N⊕H(H)–N–OH ⇌ (−H⊕/+H⊕) Ph–N=N–OH (diazotic acid)]

[⇌ (−H⊕/+H⊕)]

[Ph–N=N–O⊕H₂ →(−H₂O) Ph–N≡N⊕ Cl⊖]

═══════════════

General Reaction *(using an activated aromatic ring)*

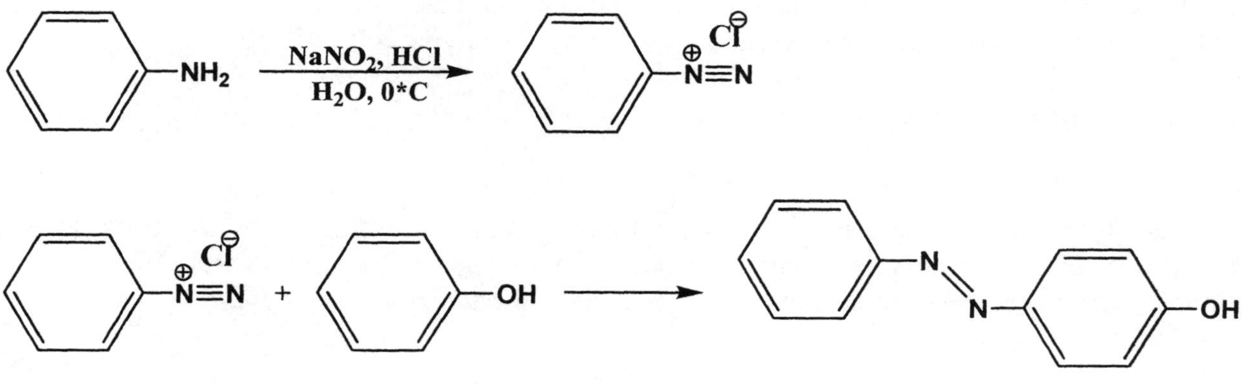

[PhNH₂ →(NaNO₂, HCl / H₂O, 0*C) Ph–N≡N⊕ Cl⊖]

[Ph–N≡N⊕ Cl⊖ + PhOH → Ph–N=N–C₆H₄–OH]

General Coupling Reaction Mechanism *(with aromatic rings)*

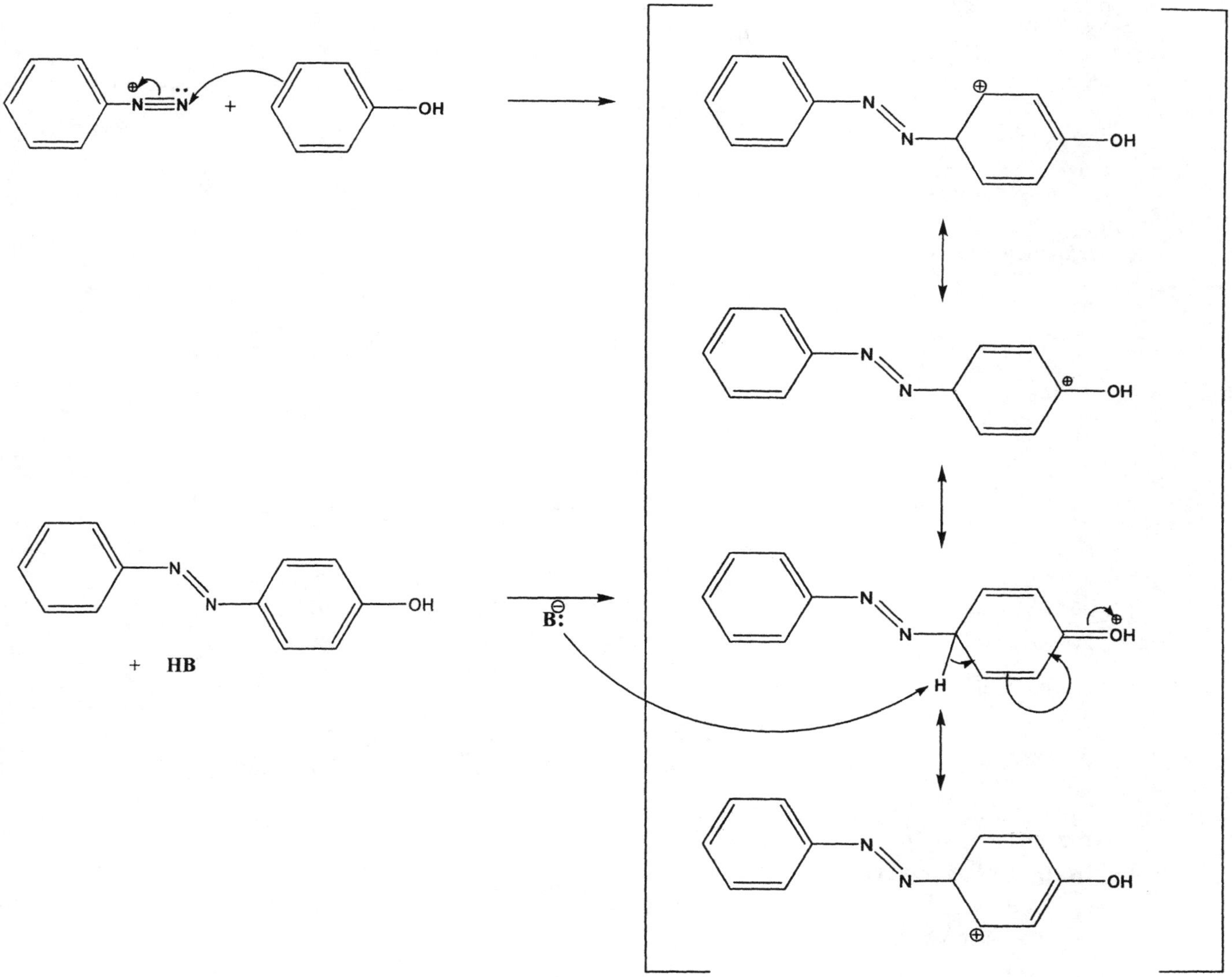

General Coupling Reaction Mechanism *(with activated methylene compounds)*
Step 1: A base is used to deprotonate the activated methylene compound.

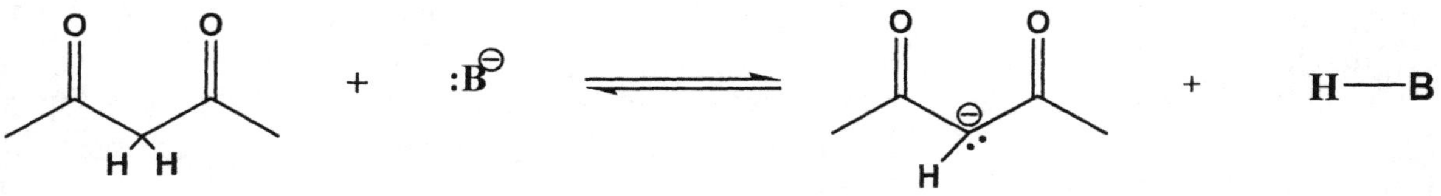

enolate anion

**NOTE: The methylene group is "activated" because the hydrogens are *very* acidic.

Step 2: The electron-rich nucleophile attacks the electrophilic diazonium salt.

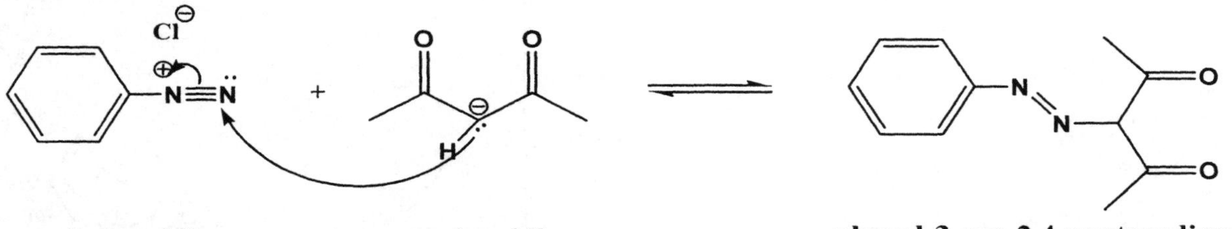

electrophile   nucleophile   phenyl-3-azo-2,4-pentanedione

Examples of Dye-Coupling Reagents:

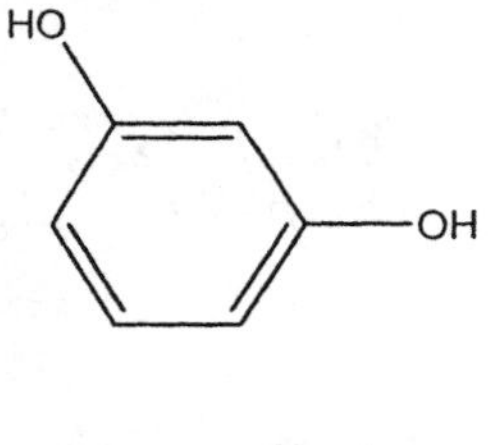

resorcinol 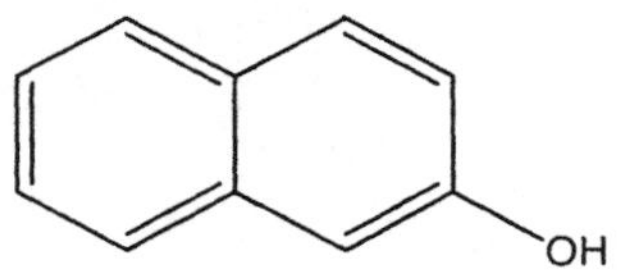   β–naphthol 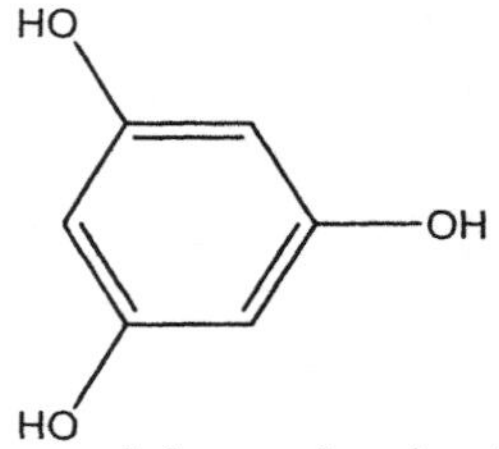   phloroglucinol

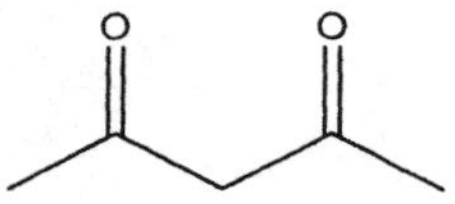

2,4-pentanedione 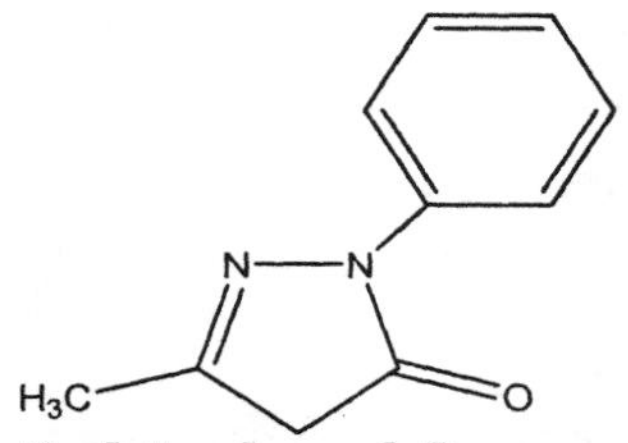   3-methyl-1-phenyl-2-pyrazolin-5-one

Procedure
Formation of Azo-Dyes:
1. In a 50 mL Erlenmeyer flask, dissolve 100 mg of *p*-toluenesulfonic acid in a 12 mL of a 1:1 ratio of acetone and ethanol.
2. To this solution, add 100 mg of 4-diazo N, N-diethylaniline fluoroborate.
3. Obtain 5 test tubes and label them with their respective dye-coupling reagents. Place 20 drops of the solution just made into each test tube. A different dye-coupler will be added to each test tube.
4. Use 1 or 2 crystals per test tube for the solid dye-couplers: resorcinol, phloroglucinol, β-naphthol, and 3-methyl-1-phenyl-2-pyrazolin-5-one. For the liquid reagent, 2, 4-pentanedione, use 2 drops of solution.
5. All solutions should appear yellow at this point in the experiment. The dye-coupler reaction cannot occur in the presence of the *p*-toluenesulfonic acid. A base must be added to initiate the occurrence of the reaction.
6. To each of the test tubes containing the diazonium salt and dye-coupler, add 1 or 2 drops of NH_3.

7. Record any color changes or other physical changes for each or the dye-couplers in your laboratory notebook.

Diazo Imaging:
1. Obtain 5 clean, dry test tubes and label them with their respective dye-coupling reagents.
2. Using the solution created in the formation of the azo-dyes (in the 50 mL Erlenmeyer flask), place 20 drops of the solution into each test tube.
3. Add the dye-couplers to their respectively labeled test tubes as before. **NOTE:** *DO NOT ADD BASE (NH$_3$)!!!*
4. All solutions should appear yellow at this point in the experiment.
5. Obtain a piece of paper and a transparency film (sheet) from your instructor. Using a black permanent marker, create a design on the transparency film. Allow this to dry completely.
6. Using a paint brush, coat the paper with the different dye solutions. Allow the dye solutions to dry completely.
7. Place the transparency film over the paper, as per your instructor's demonstration, and secure with a paperclip.
8. Expose the clipped paper and film to a flood lamp for a minimum of 25 minutes.
9. Following exposure, remove the film from the coated paper and expose it to NH$_3$ vapors by hanging it in the large development beaker located in the fume hood.
10. Record all observation in your laboratory notebook.

References
Organic Chemistry 7th ed.; John McMurry.
Introduction to Organic Laboratory Techniques: A Microscale Approach 3rd ed.; Pavia, Lampman, Kriz, and Engel.

Safety
1. **Take precaution when using the reagents; AVOID inhalation of the NH$_3$ gas/fumes.**
2. **Clean (using soap for glassware) and rinse (using distilled water for final rinse) ALL glassware and utensils PRIOR to returning them to your lab drawer.**
3. **If exposure to ANY of the reagents occurs, be sure to wash hands and any other affected areas thoroughly with plenty of soap and water.**

Disposal
Dispose of the liquid reagents used in the "Organic Liquid Wastes" receptacle.
Dispose of the dyes formed in the "Organic Liquid Wastes" receptacle.
Dispose of any disposable test tubes and/or pipettes in the "broken glass" bin.

Postlab Questions
1. Draw the structures of ALL the azo dyes you prepared.
2. Provide the complete electrophilic aromatic substitution mechanism for the dye-coupling reaction between benzenediazonium chloride and β-naphthol. Include all resonance structures for the intermediates formed in the reaction and the final products.
3. What roles do the ammonia and halogen lamp play in the developing of the colorful image formed?

Preparation of an α,β-Unsaturated Ketone— Exp#20

Background
Compounds with a carbonyl functional group can act as either an electrophile *or* a nucleophile depending on the reaction conditions.

Theory
An α,β-unsaturated ketone can be formed via a Michael Addition and Aldol Condensation Reaction. A strong base, NaOH, is used as a catalyst to drive the reaction toward completion in the formation of 6-ethoxycarbonyl-3,5-diphenyl-2-cyclohexenone.

Terminology
1. *Tautomers*: A pair of constitutional isomers that convert into each other (e.g., keto form to enol form) as shown below.
2. *Enolization*: The process of converting a carbonyl compound from the "keto" form to the "enol" form.

Example of Enolization:

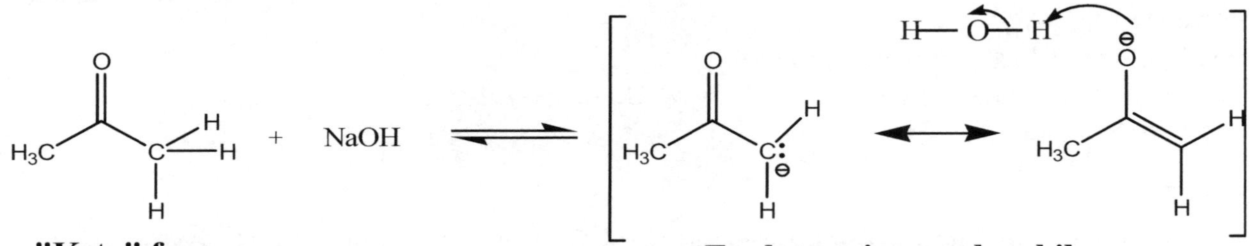

"Keto" form → Enolate anion-nucleophile → "Enol" form

Electrophilic Potential of Carbonyl Compound:

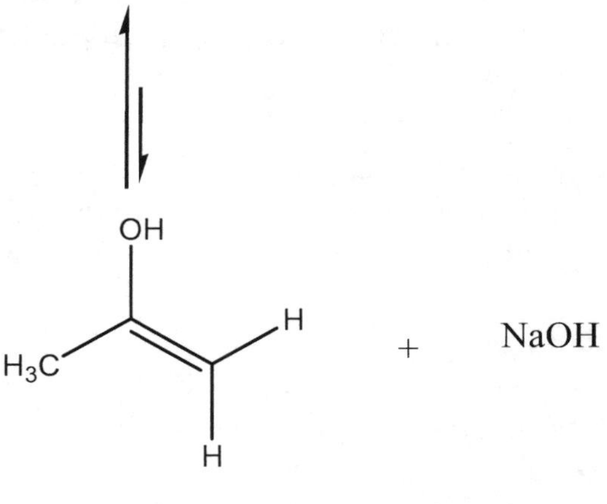

**(+) = electrophilic sites

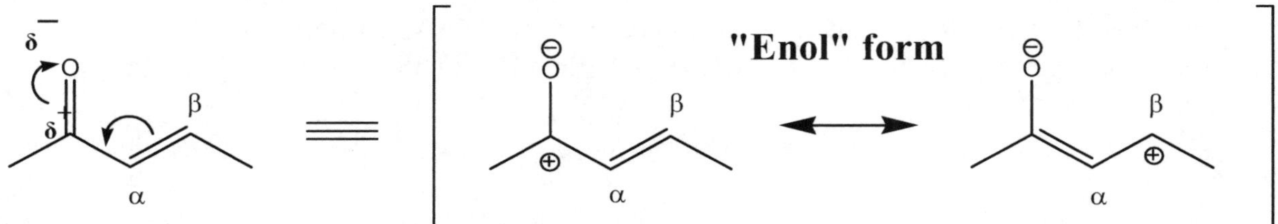

Specific Mechanism

Step 1—Base catalyzed formation of an enolate anion:

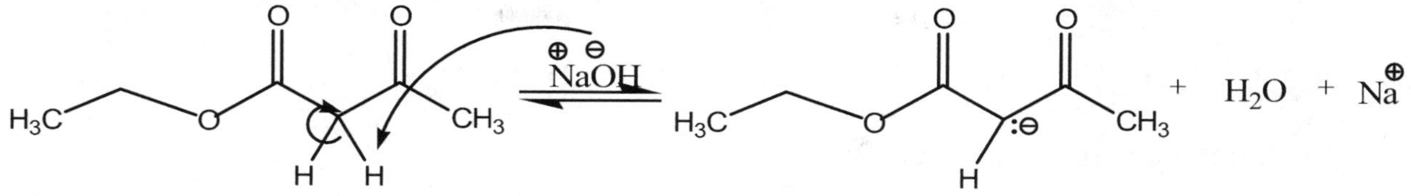
H3C
CH3
H
NaOH
H2O
Na

Enolate anion

Step 2—Michael Addition:

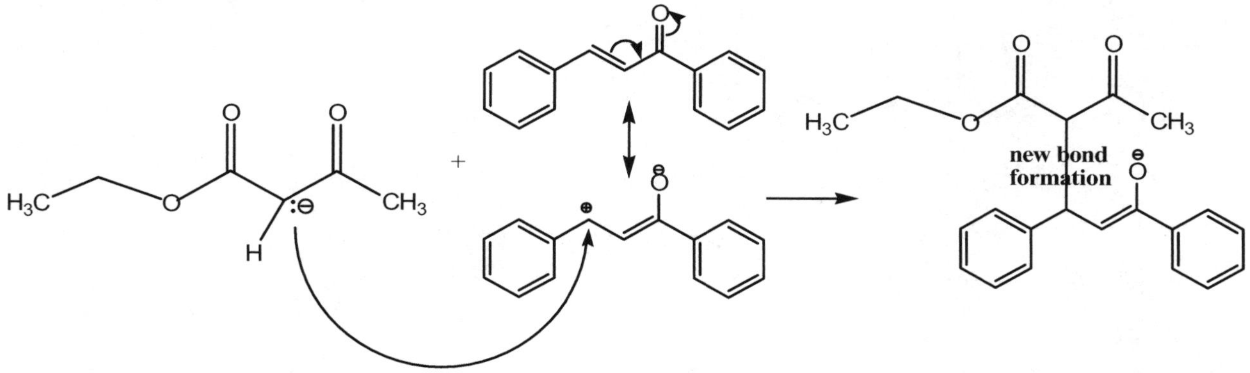
H3C
CH3
H
new bond formation

Step 3—Tautomerization:

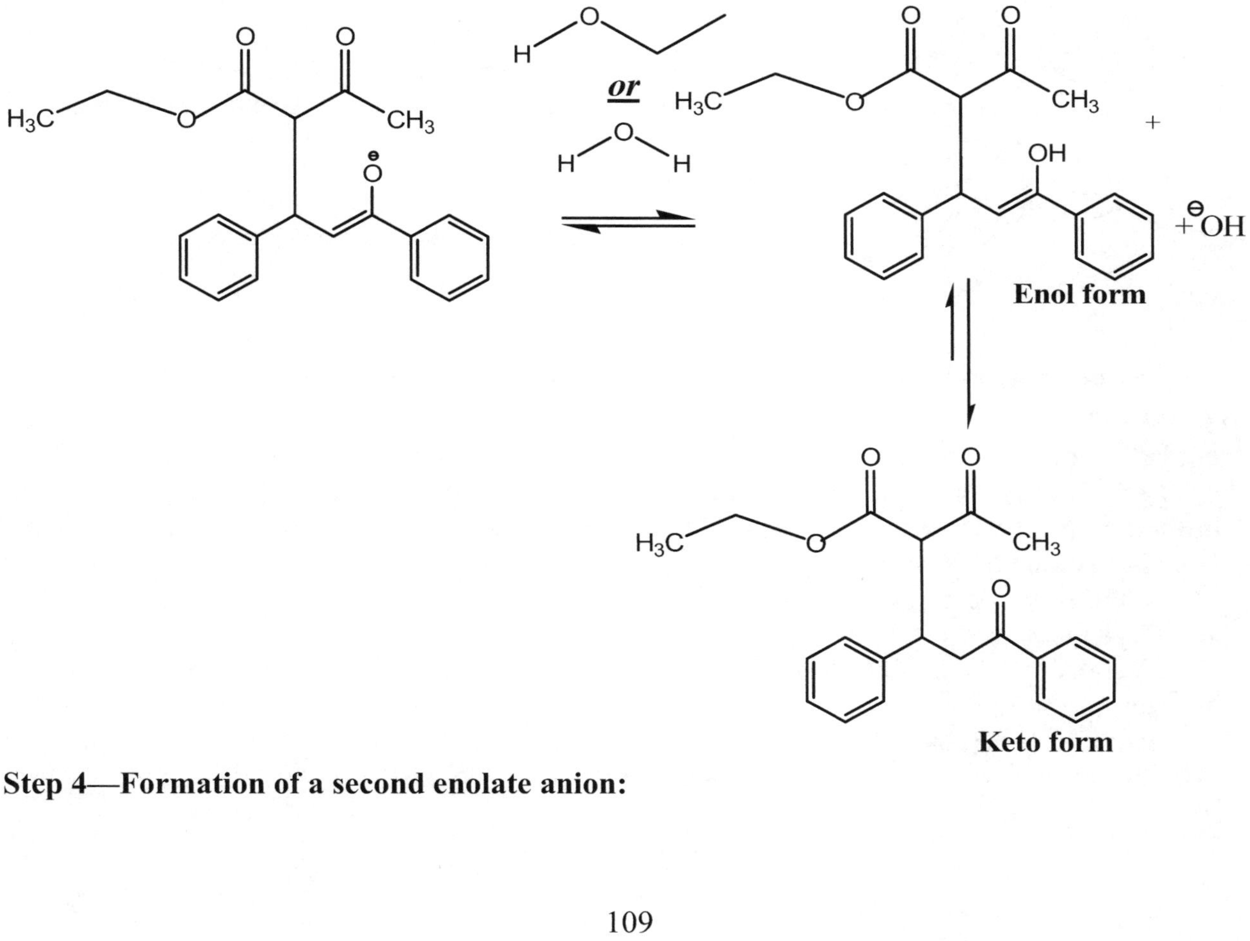
H3C
CH3
or
OH
Enol form
Keto form

Step 4—Formation of a second enolate anion:

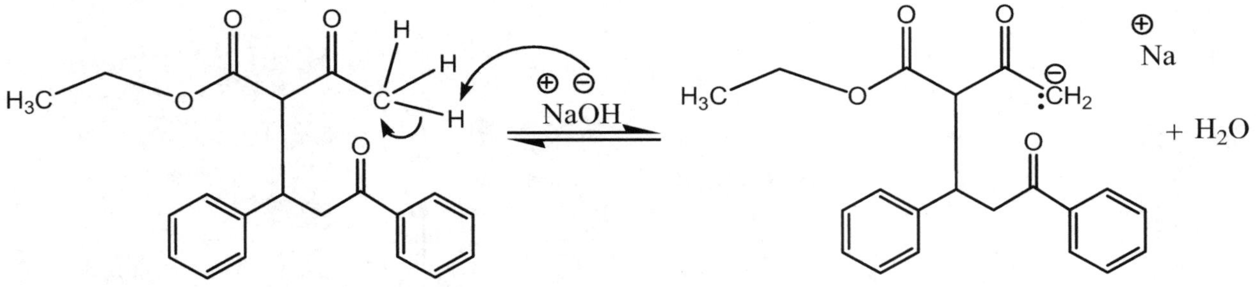

Step 5—Aldol Condensation (part I):

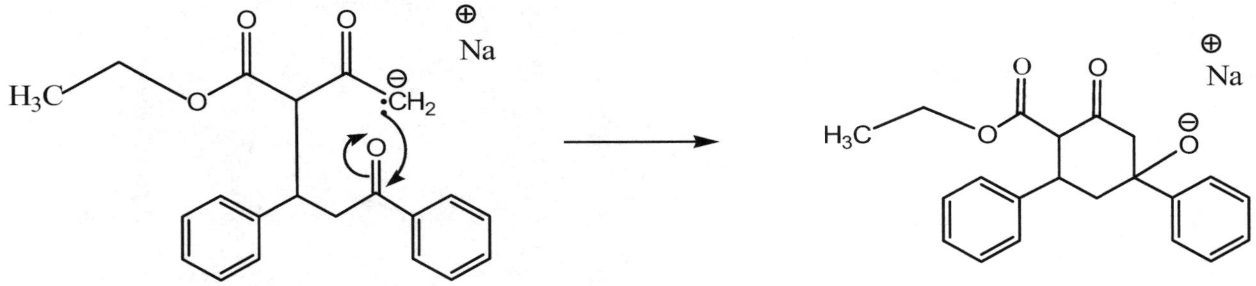

Step 6—Aldol Condensation (part II):

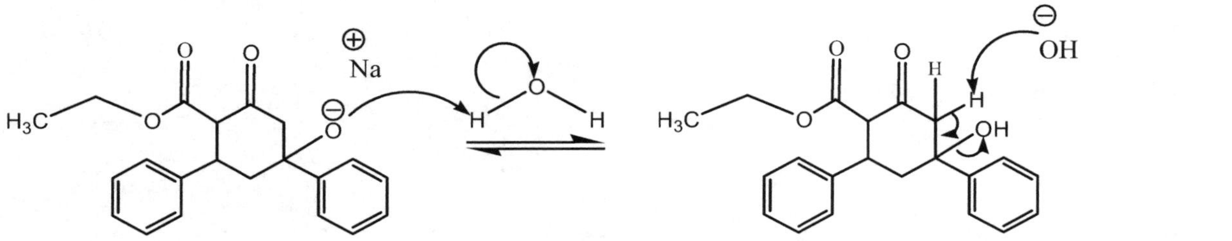

E2 like mechanism with a loss of H₂O

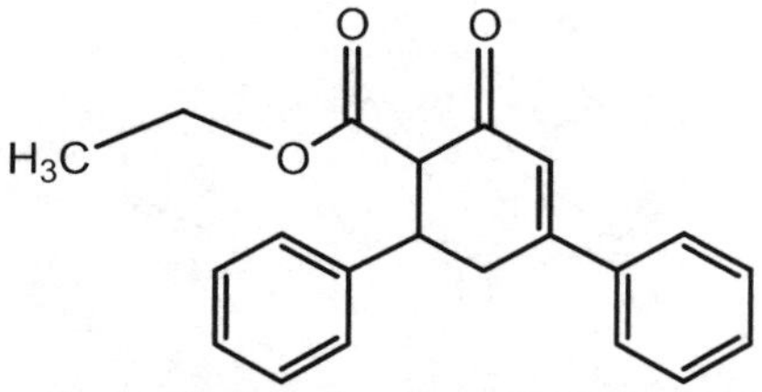

Final product: 6-ethoxycarbonyl-3,5-diphenyl-2-cyclohexenone

Procedure

Preparation of crude 6-ethoxycarbonyl-3,5-diphenyl-2-cyclohexenone:

<u>NOTE</u>: Because a one hour reflux is required, you should start the experiment immediately at the very beginning of the laboratory period. During the reflux period, your instructor may lecture.

1. Weigh out 1.15 mmol of ethylacetoacetate into a 10 mL round-bottom flask using the Tare method.
2. Weigh out 1.15 mmol of finely ground *trans*-chalcone to the same 10 mL round-bottom flask.
3. Add 5 mL of **absolute ethanol** solution to the flask and swirl until either all or most of the solids have dissolved. Place one boiling chip in the flask.

4. Add 0.25 mL (~8 drops using a short-stem Pasteur pipette or micropipette) of a 2.2 M NaOH to the mixture in the flask.
5. Clamp the flask to a ring stand and attach a water-cooled condenser to the flask as shown in Figure 25. **NOTE: Prior to assembly, be sure to use silicon grease to lubricate the "ground" glass portions of the reaction apparatus. *THIS IS NECESSARY TO PREVENT FUSION OF THE PARTS OF YOUR APPARATUS!***

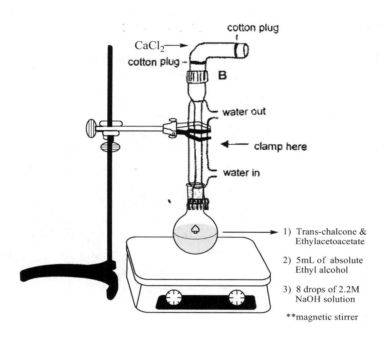

Figure 25: Demonstration of reaction apparatus

6. Heat the reaction mixture on a hot plate until a gentle boil is reached.
7. Once the mixture begins to reflux, continue to heat for at least one hour. **NOTE:** *The mixture may become cloudy during this reflux period.*

Isolation of the crude product:
1. Following the one hour of reflux, let your reaction cool to room temperature, then add 2 mL of water.
2. Using a glass stirring rod, vigorously scratch the sides of the reaction vessel to induce crystallization of your product.
3. Place the reaction flask in an ice bath for at least 20 minutes in order to completely crystallize the product.
4. You should periodically scratch the sides of the reaction vessel to aid in the crystallization process.
5. Using a Hirsch funnel and side arm tube, vacuum filter the crystals. Use 1 mL of ice cold distilled water to help in the transfer.

6. Following the transfer, rinse the flask with 0.5 mL of cold 95% ethanol to transfer the remaining solid crystals from the flask to the funnel.
7. Spread the crystals out on a large piece of filter paper to help dry them out.
8. Once the crystals are completely dry, place the filter paper on a watch glass and let the crystals air dry in your drawer until the next laboratory period.

STOP HERE!! (Complete the procedure next class period.)

Removal of catalyst:
1. Carefully transfer your product to a clean, dry centrifuge test tube.
2. Add 1.5 mL of reagent grade acetone to the centrifuge tube and stir the mixture with a spatula. ****NOTE:** *Most of the solid will dissolve in the acetone; however, the NaOH catalyst and salt by-products will be insoluble.*
3. Let the solids settle, make sure to leave as much of the solid behind as possible. The solution should be very cloudy, but centrifugation of the mixture will help separate the rest of the solid from the liquid.
4. Centrifuge the acetone mixture for approximately 2 minutes. ****NOTE:** *Be sure to balance the centrifuge with either another group's sample or a centrifuge filled with an equivalent of water.*
5. Preweigh a clean, dry test tube containing a boiling chip. Carefully transfer the clear portion of the acetone solution to the preweighed tube using a Pasteur pipette.
6. Prepare a hot water bath of approximately 60 °C (not boiling). Use this hot water bath to carefully evaporate the remaining acetone from the product. ****NOTE:** *Be careful to avoid spattering of the solution.*
7. The evaporated material may contain an oily residue as it cools. Use a spatula to aid in inducing crystallization.
8. Reweigh the test tube and record the weight of your crude product in your laboratory notebook.

Purification of the product by recrystallization:
1. Add a boiling chip to a 25 mL Erlenmeyer flask, then add 5 mL of a 95% ethanol solution. Set your hot plate at ~3 and heat the solution to boiling. Add 5 mL of ethanol to a small beaker and place it in an ice water bath—set aside.
2. Place the test tube containing the crude product in a hot water bath of ~60 °C. Continuously add the heated ethanol solution to the crude product using a filter-tipped Pasteur pipette. Add the ethanol until the crude product is dissolved.
3. Let the crude product solution sit undisturbed for ~8 minutes. Crystals should begin to appear. ****NOTE:** *If crystals do not form, consult your lab instructor.*
4. Place the product solution in an ice water bath for a ***minimum of 15 minutes***.
5. Collect the purified product by vacuum filtration using a Hirsch funnel.
6. Use 0.5 mL of ice-cold ethanol (set aside from before) to help aid in the transfer of the product to the funnel. Let air pass through the funnel for ~8 minutes at high vacuum suction and until no water drips from the stem.
7. Spread your crystals out on a preweighed watch glass and let stand at room temperature for at least 20 minutes.
8. The product should now be completely dry and the melting point can be measured. Record the results in your lab notebook.

Characterization
1. Obtain the melting point of the pure product you have isolated.
2. Obtain an IR spectrum of your product by preparing a KBr pellet. Identify **ALL** functional groups and assign **ALL** relative vibrational peaks.
3. The ^{1}H NMR spectrum of the product will be provided for you, assign **ALL** peaks in the spectrum.
4. For your purified product, calculate the experimental percent yield.

References
Organic Chemistry 7th ed.; John McMurry.
Introduction to Organic Laboratory Techniques: A Microscale Approach 3rd ed.; Pavia, Lampman, Kriz, and Engel.

Safety
1. **Take precaution when handling the NaOH because it is <u>CORROSIVE</u>; if exposure occurs, rinse the affected areas with plenty of water. Following rinsing, wash the affected areas with plenty of soap and water.**
2. **Take precaution when handling <u>HOT</u> objects.**
3. **Turn <u>OFF</u> any burners and/or hot plates when not in use.**
4. **Clean (using soap for glassware) and rinse (using distilled water for final rinse) <u>ALL</u> glassware and utensils <u>PRIOR</u> to returning them to your lab drawer.**
5. **If exposure to <u>ANY</u> of the reagents occurs, be sure to wash hands and any other affected areas thoroughly with plenty of soap and water.**

Disposal
Dispose of excess liquid reagents in the "Organic Liquid Wastes" receptacle.
Dispose of the product in the "Organic Solid Wastes" receptacle.
Dispose of any disposable pipettes and/or test tubes in the "broken glass" bin.

Postlab Questions
1. The following questions are in reference to the Michael addition and Aldol condensation reaction mechanisms.
    a. Draw the nucleophilic species involved in the Michael addition portion of the reaction.
    b. Draw the electrophilic species involved in the Michael addition portion of the reaction.
    c. For the intramolecular aldol condensation reaction, draw the molecular structure of the reactant and identify the nucleophilic and electrophilic carbons involved by circling them.
2. *Trans*-chalcone is classified as an α,β-unsaturated ketone, using an aldol condensation reaction, show how this compound can be formed.
3. Why is water added to the reaction mixture during the isolation of the crude product?

Condensation Polymers or Step-Growth Polymers: Nylon and Polyester—Exp#21

Background
Last semester, you prepared addition polymer or chain-growth polymer, now you will synthesize condensation polymer. These polymers, as their old name suggests, are prepared by condensation reactions in which monomeric subunits are joined through intermolecular eliminations of small molecules such as water or alcohols. The monomers used in this kind of polymerization are bifunctional (**_DEFINITION_**: each monomer bears two different functional groups or the same functional group; also each monomer might have three different functional groups or the same or more). Among the most important condensation polymers are polyamides, polyesters, polyurethanes, and phenol-formaldehyde resins. **NOTE:** polyurethanes are formed by step-growth polymerization but no small molecule is eliminated.

Silk and wool are two examples of polyamides. These make up a unique family called proteins. The repeating units of proteins are derived from α-amino acids and that these subunits are joined by amide linkages as shown on pages 99 and 100. Therefore, proteins are polyamides.

Applications
Nylon can be drawn out in fibers to produce fabrics, carpeting, and tire cords, and it can be molded to form many objects; and the polyester can be used in paint-lacquers, beverage bottles, films and similar products.

In this experiment, you will synthesize the polymer nylon 6,6 and a polyester: poly(propylenephthalate) and compare their physical properties with each other.

Nylon by Interfacial Polymerization
In today's lab experiment a diacid chloride dissolved in cyclohexane is carefully floated on top of a solution of a diamine dissolved in water. Where the two solutions come in contact (the interface), a nucleophilic acyl substitution reaction occurs to form a film of polyamide. The reaction stops there unless the polyamide is removed. The polymer can be picked up with a wire hook and continuously removed in the form of a rope. As the diamine diffuses into the organic layer, reaction occurs immediately to give the insoluble polymer. The HCl produced reacts with the sodium hydroxide in the aqueous layer. The acid chloride does not hydrolyze before reacting with the amine because it is not very soluble in water.

Synthesis of Nylon in Industry
One of the most important nylons, called **_nylon-6,6_** can be prepared from the six-carbondicarboxylic acid, adipic acid and the six-carbon diamine, hexamethylenediamine. In the commercial process these two compounds are allowed to react in equimolar proportions in order to produce a 1:1 salt.

hexamethylenediamine + **adipic acid**

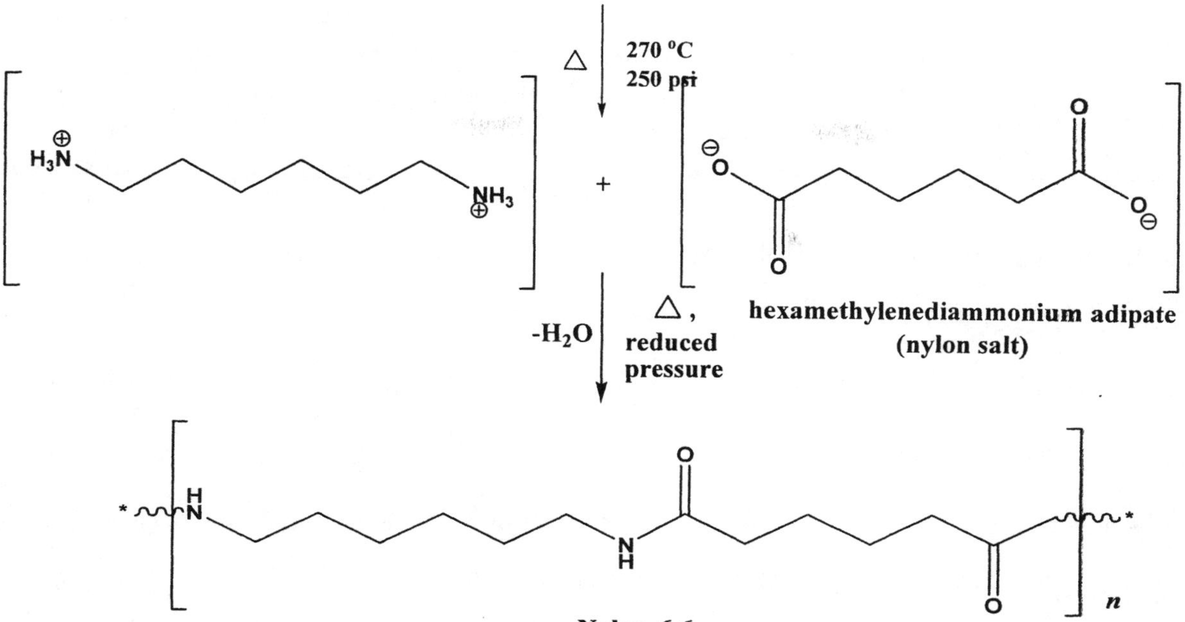

[Reaction scheme: hexamethylenediammonium adipate (nylon salt) formed from protonated hexamethylenediamine and adipate dianion, at 270 °C, 250 psi, Δ; then with −H$_2$O, Δ, reduced pressure → Nylon-6,6]

General Reaction

$$n\ F_1\text{—M—}F_2 \longrightarrow n\ F\text{—M—}F$$

↑ ↑
bifunctional monomer

F_1 = first reactive bifunctional monomer group 1.
F_2 = second (different) reactive bifunctional monomer group 2.
F = a new functional group (formed via reaction).

For example,

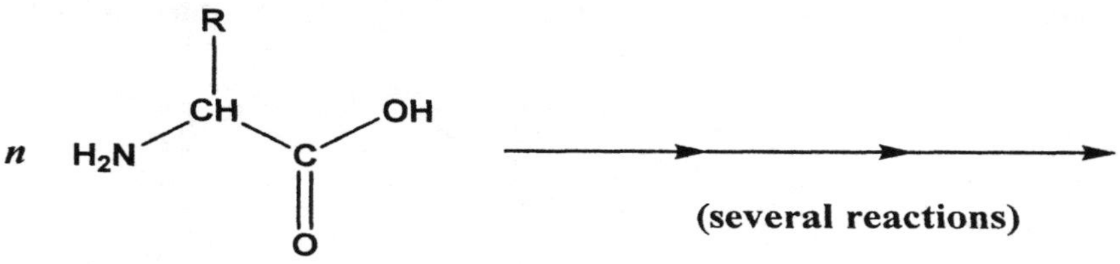

n H$_2$N–CH(R)–C(=O)OH an α-amino acid $\xrightarrow{\text{(several reactions)}}$

[polypeptide chain with repeating —NH—CH(R)—C(=O)— units]

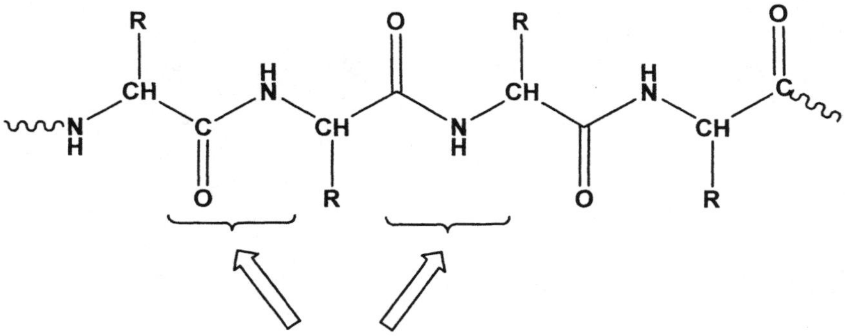

amide linkage

115

NOTE: This is an example of a portion of a polyamide chain as it might occur in a protein.

General Reaction for Condensation Polymerization (Nylon)

$F_1 - M_1 - F_1 \ + \ F_2 - M_2 - F_2 \longrightarrow \ \sim[M_1 - F - M_2]\sim \ + $ small molecule by-product

Specific Reaction for Condensation Polymerization (Nylon)

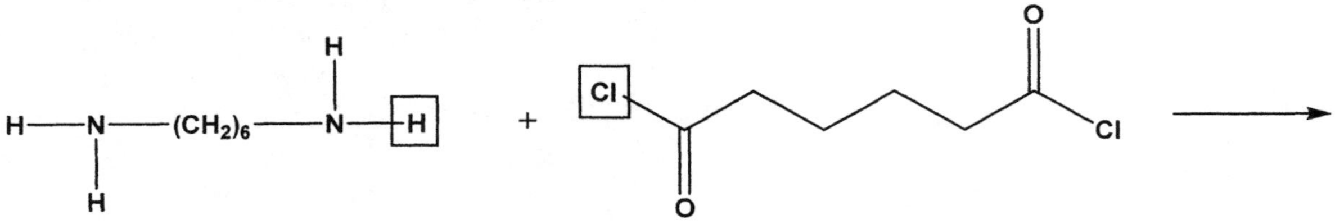

hexamethylenediamine + adipoylchloride ⟶

> **HCl** is released as a small molecule by-product.

new bond

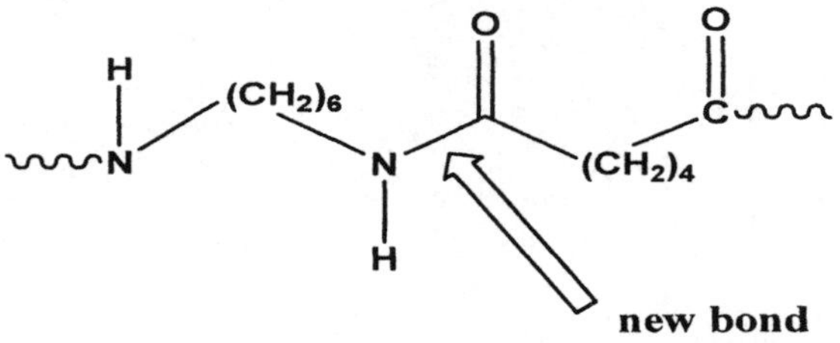

Specific Mechanism for Condensation Polymerization (Nylon)

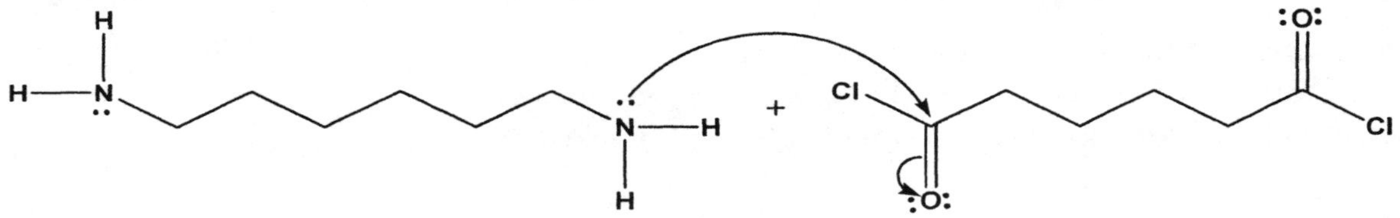

hexamethylene diamine + adipoyl chloride

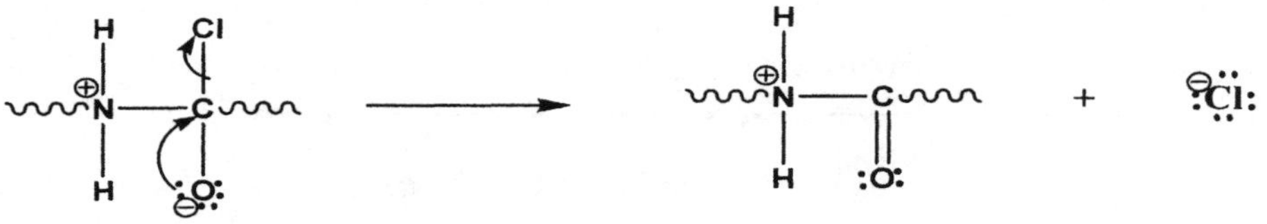

NOTE: a portion of the molecule showing the specific mechanism

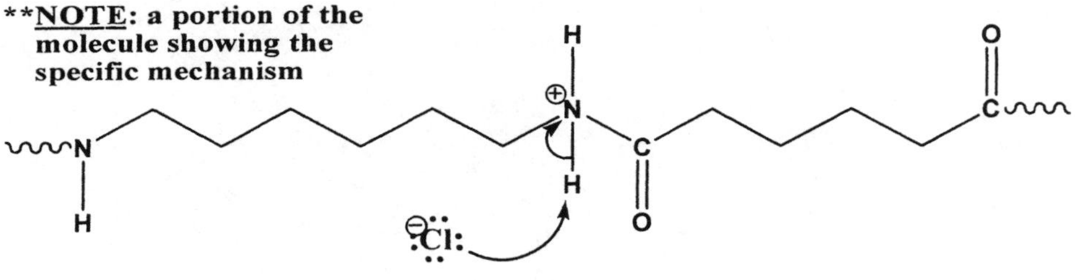

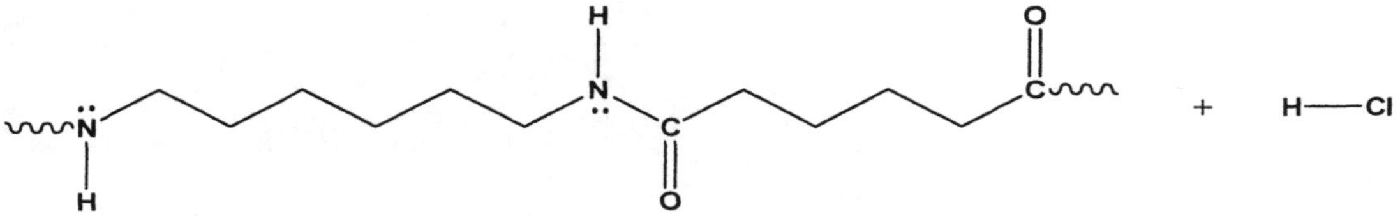

+ H—Cl

nylon-6,6

General Reaction for Condensation Polymerization (Polyester)

For example,

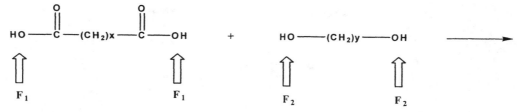

F_1 = reactive carboxylic acid functional group.
F_2 = second reactive functional group.
$x = 4, 6, 8, \ldots$ etc.
$y = 2, 3, 4, \ldots$ etc.

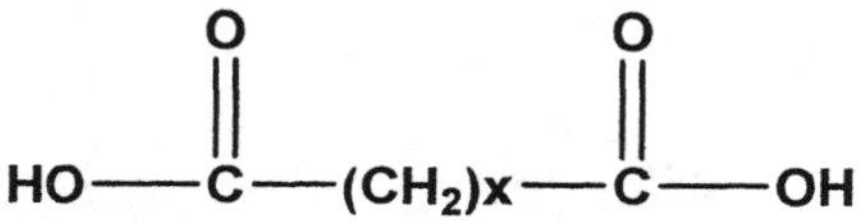

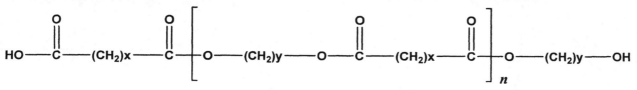

Bifunctional monomer #1 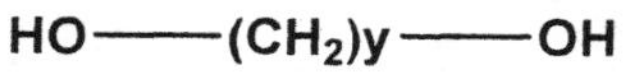 Bifunctional monomer #2

repeating unit in a "linear" polyester

$+ (n + 2) H_2O$

Specific Reaction for Condensation Polymerization (Polyester)

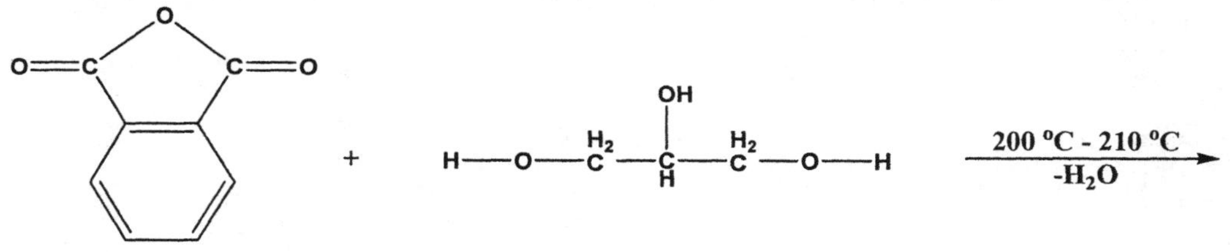

phthalic anhydride + glycerol $\xrightarrow[-H_2O]{200\ °C - 210\ °C}$

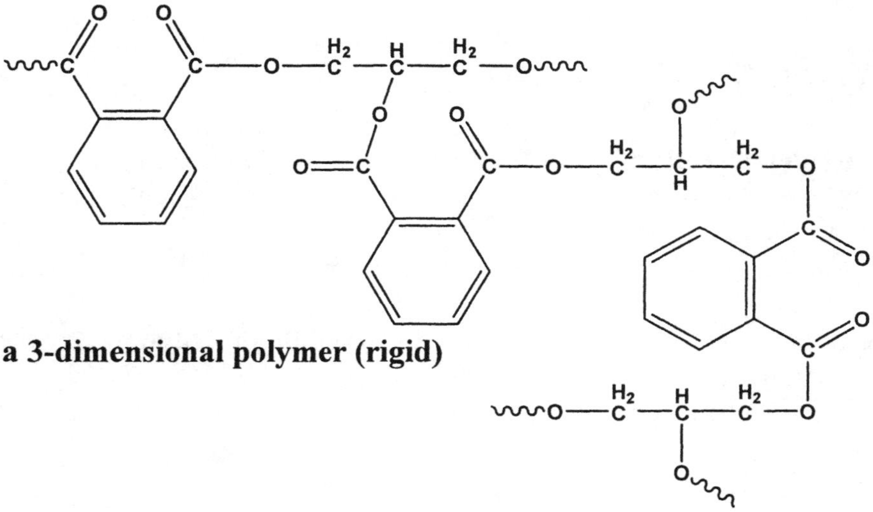

Cross-linked to form a 3-dimensional polymer (rigid)

Specific Mechanism for Condensation Polymerization (Polyester)

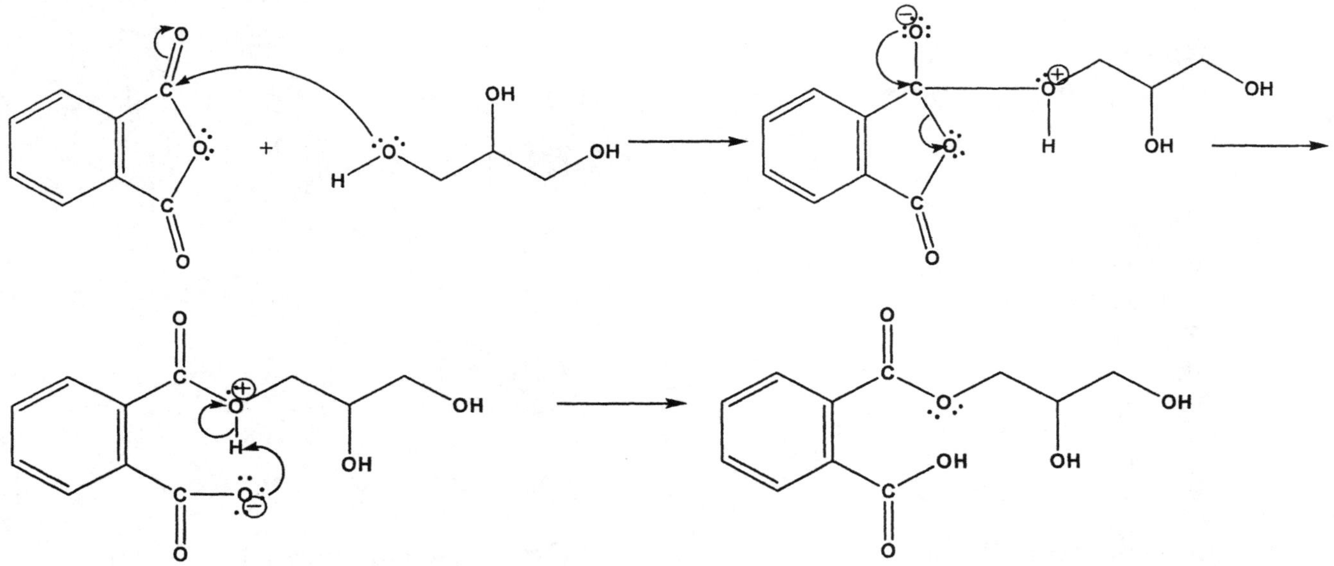

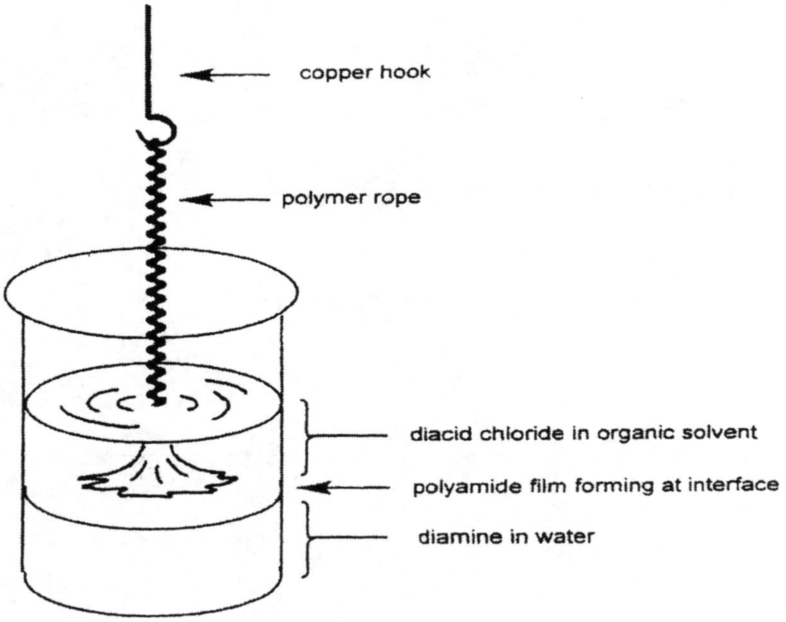

Figure 24: Interfacial polymerization

Procedure (Polyamide Nylon-6,6)
1. Place 10 mL of a 5% aqueous solution of hexamethylene diamine and 10 drops of a 20% NaOH solution into a 50 mL beaker.
2. Measure 10 mL of the 5% adipoyl chloride in cyclohexane solution using a graduated cylinder.
3. Carefully and slowly pour the adipoyl chloride solution down the side of the beaker containing the hexamethylene diamine and NaOH.
4. A polymer film will form immediately at the interface of the two solutions.

5. Using a copper wire with a hooked-end, carefully free the polymer from the sides of the beaker. Then, hook the polymer mass in the center of the beaker and draw it up as demonstrated in Figure 24.
6. **DO NOT TOUCH** the polymer rope you have formed until you wash it thoroughly with water.
    > *CAUTION*: Adipoyl chloride and hexamethylenediamine are toxic. Avoid skin contact and do not breathe their vapors. Also avoid skin contact with the 20% sodium hydroxide. *Wear disposable gloves.*
7. While drawing the polymer up, remove ~40 cm of the product. Stretch the end of the remaining rope over a nut to be used for IR analysis.
8. Place the rope on a paper towel to dry.
9. With your spatula, gently stir the remaining polymer mixture in the beaker.
10. Allow the product to dry on the nut, then run an IR on this film. (The instructor will show you this technique during the prelab discussion.)

Procedure (Polyester)
1. Weigh 500 mg of phthalic anhydride and 25 mg of sodium acetate and place them together in a test tube.
2. Add 10 drops of glycerol. Gently heat the test tube until the solution appears to boil. From this point, continue to heat strongly over your hotplate for an additional 5 minutes.
3. Remove the test tube from the heat and allow it to cool.
4. The solution should become viscous as it cools and it should also form a solid plug within the test tube.

Polymer Characterization

Nylon-6,6
1. Polyamide film: Refer to step 10 in the procedure for polyamide nylon-6,6 synthesis.

Polyester
1. Polyester film: Add 1 mL of CH_2Cl_2 to the solid mass of polyester in the test tube.
2. Swirl your spatula to help dissolve some of the polymer in the solvent (**the majority of the polymer will *NOT* dissolve; however, enough of the polymer should dissolve into solution in order to cast a film for IR analysis**).
3. Place one drop of the dissolved polymer on a NaCl salt plate. Allow the solvent to evaporate *(consult with your instructor)*. Run an IR once your instructor has approved your sample.
4. The film can be removed from the NaCl salt plates by redissolving the polymer in CH_2Cl_2 and then wiping the residue off the plates.

References
Organic Chemistry 7th ed.; John McMurry.
Introduction to Organic Laboratory Techniques: A Microscale Approach 3rd ed.; Pavia, Lampman, Kriz, and Engel.

Safety
1. **Adipoyl chloride and hexamethylenediamine are toxic. Avoid skin contact and do not breathe their vapors. Also avoid skin contact with the 20% sodium hydroxide. Wear disposable gloves.**
2. **Take precaution when handling HOT objects.**
3. **Turn OFF any burners and/or hot plates when not in use.**
4. **Clean (using soap for glassware) and rinse (using distilled water for final rinse) ALL glassware and utensils PRIOR to returning them to your lab drawer.**
5. **If exposure to ANY of the reagents occurs, be sure to wash hands and any other affected areas thoroughly with plenty of soap and water.**

Disposal
After pulling out the nylon fiber about 40 cm, stir the mixture vigorously to cause the nylon to precipitate. Decant the aqueous and organic layer in the organic liquid waste and dispose of the nylon bulb after squeezing in the "Organic Solid Wastes" receptacle.

Postlab Questions
1. Draw the polymer resulting from the condensation reactions of the following monomers:
   (a) HOOC–CH₂–CH₂–OH $\xrightarrow{\Delta}$

   (b) HOOC–C₆H₄–COOH + H₂N–CH₂–CH(NH₂)–CH₂–NH₂ $\xrightarrow{\Delta}$

2. Draw the two bifunctional monomers that were used to synthesize the following polymers:
   (a) [–O–C₆H₄–C(=O)–]$_n$

   (b) Nylon-6,10

3. Polyamides are condensation polymers formed by the reaction of a diacid anhydride and a diamine. Initially, an intermediate poly(amic-acid) is formed followed by loss of water to form a cyclic imide structure. Draw the intermediate poly(amic-acid) and resulting polyimide formed by the reaction of pyromellitic dianhydride and hexamethylene diamine.

   pyromellitic dianhydride + H₂N–(CH₂)₆–NH₂ $\xrightarrow{\text{DMF}}$ poly(amic-acid) ? $\xrightarrow[\text{DMF}]{\Delta, -H_2O}$ polyimide

4. Explain why polymers can be cast into a film while the monomers they are prepared from cannot.

Solubility Chart

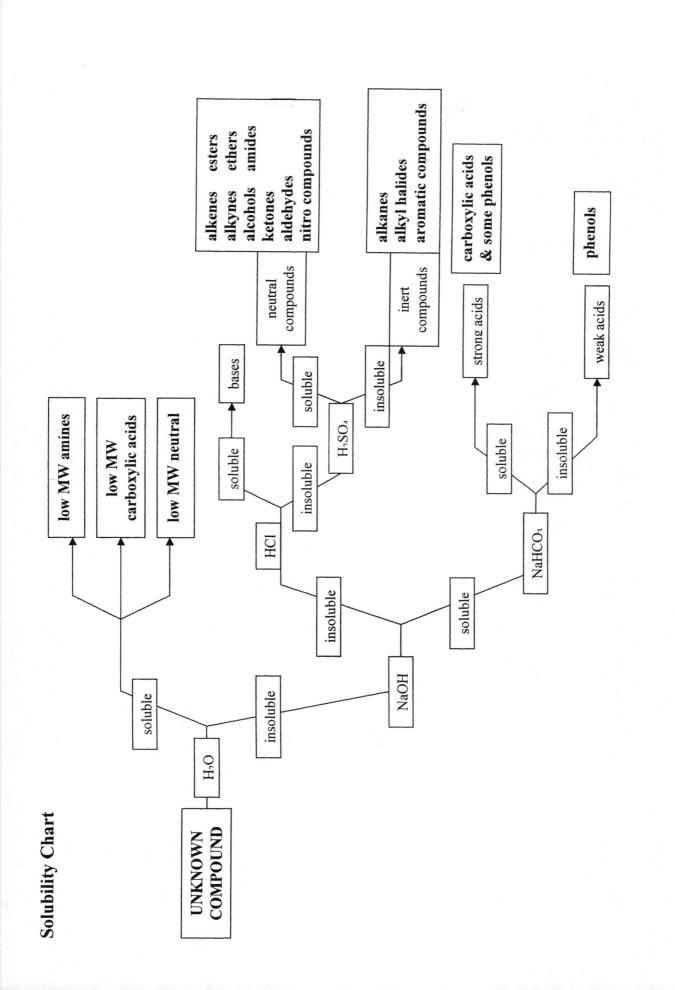